The Genesis of Innovation

NEW HORIZONS IN THE ECONOMICS OF INNOVATION

Founding Editor: Christopher Freeman, *Emeritus Professor of Science Policy, SPRU – Science and Technology Policy Research, University of Sussex, UK*

Technical innovation is vital to the competitive performance of firms and of nations and for the sustained growth of the world economy. The economics of innovation is an area that has expanded dramatically in recent years and this major series, edited by one of the most distinguished scholars in the field, contributes to the debate and advances in research in this most important area.

The main emphasis is on the development and application of new ideas. The series provides a forum for original research in technology, innovation systems and management, industrial organization, technological collaboration, knowledge and innovation, research and development, evolutionary theory and industrial strategy. International in its approach, the series includes some of the best theoretical and empirical work from both well-established researchers and the new generation of scholars.

Titles in the series include:

Embracing the Knowledge Economy
The Dynamic Transformation of the Finnish Innovation System
Edited by Gerd Schienstock

The Dynamics of Innovation in Eastern Europe
Lessons from Estonia
Per Högselius

Technology and the Decline in Demand for Unskilled Labour
A Theoretical Analysis of the US and European Labour Markets
Mark Sanders

Innovation and Institutions
A Multidisciplinary Review of the Study of Innovation Systems
Edited by Steven Casper and Frans van Waarden

Innovation Strategies in Interdependent States
Essays on Smaller Nations, Regions and Cities in a Globalized World
John de la Mothe

Internationalizing the Internet
The Co-evolution of Influence and Technology
Byung-Keun Kim

Asia's Innovation Systems in Transition
Edited by Bengt-Åke Lundvall, Patarapong Intarakumnerd and Jan Vang-Lauridsen

National Innovation, Indicators and Policy
Edited by Louise Earl and Fred Gault

Innovation and the Creative Process
Towards Innovation with Care
Edited by Lars Fuglsang

The Genesis of Innovation
Systemic Linkages Between Knowledge and the Market
Edited by Blandine Laperche, Dimitri Uzunidis and Nick von Tunzelmann

The Genesis of Innovation

Systemic Linkages Between Knowledge and the Market

Edited by

Blandine Laperche

Research Unit on Industry and Innovation, University of Littoral Côte d'Opale, France

Dimitri Uzunidis

Research Unit on Industry and Innovation, University of Littoral Côte d'Opale, France

Nick von Tunzelmann

Professor in the Economics of Science and Technology, SPRU – Science and Technology Policy Research, University of Sussex, UK

NEW HORIZONS IN THE ECONOMICS OF INNOVATION

Edward Elgar

Cheltenham, UK • Northampton, MA, USA

Published by
Edward Elgar Publishing Limited
The Lypiatts
15 Lansdown Road
Glos GL50 2JA
UK

Edward Elgar Publishing, Inc.
William Pratt House
9 Dewey Court
Northampton
Massachusetts 01060
USA

A catalogue record for this book
is available from the British Library

Library of Congress Control Number: 2008927977

ISBN 978 1 84720 593 3

Printed and bound in Great Britain by MPG Books Ltd, Bodmin, Cornwall

Contents

PART III FROM KNOWLEDGE TO MARKET: SYSTEMIC LINKS AT THE LOCAL LEVEL

Figures

Tables

Boxes

Contributors

Marie-Pierre Bès is Associate Professor of Economics, Director of Research at the Interdisciplinary Research Center for Urban Studies and Sociology, CNRS/UTM & ENSICA, France.

Sophie Boutillier is Associate Professor of Economics, Director of Research at the Research Unit on Industry and Innovation, University of Littoral Côte d'Opale, France.

Michel Callois is a PhD student, GREDEG-DEMOS, University of Nice Sophia Antipolis, France.

Elena Castro-Martínez works as a scientist at INGENIO, a joint institute of the Spanish Council for Scientific Research (CSIC) and the Polytechnic University of Valencia, Spain.

Ignacio Fernández-de-Lucio is research scientist at INGENIO, a joint institute of the Spanish Council for Scientific Research (CSIC) and the Polytechnic University of Valencia, Spain.

Alfredo Ilardi is former Director, Head of the Collection of Laws Section of the World Intellectual Property Organization (WIPO), Switzerland.

Marc Isabelle is Doctor in Economics, and researcher at Commissariat à l'Energie Atomique and IMRI, University of Paris-Dauphine, France.

Robyn Keast is Research Fellow, School of Management, Queensland University of Technology, Brisbane, Australia.

Stefan Kesting is Senior Lecturer in Economics, Faculty of Business, Auckland University of Technology, New Zealand.

Blandine Laperche is Associate Professor of Economics, Director of Research at the Research Unit on Industry and Innovation, University of Littoral Côte d'Opale, France.

Kavoos Mohannak is Lecturer, School of Management, Queensland University of Technology, Brisbane, Australia.

Jordi Molas-Gallart is Research Professor at INGENIO, a joint institute of the Spanish Council for Scientific Research (CSIC) and the Polytechnic University of Valencia, Spain.

Claude Serfati is Associate Professor of Economics, Director of Research at lab.C3ED at the University of Versailles-Saint Quentin-en-Yvelines, France.

Paul Sommers is Professor, Institute of Public Service and Albers School of Business, Seattle University, United States of America.

Dimitri Uzunidis is Director of the Research Unit on Industry and Innovation (Lab.RII), University of Littoral Côte d'Opale, France and Professor at Technical University of Crete, Greece.

Nick Von Tunzelmann is Professor of Economics at SPRU, University of Sussex, United Kingdom.

Introduction: knowledge, innovation systems and the role of power

Blandine Laperche, Dimitri Uzunidis and Nick von Tunzelmann

Knowledge and innovation are the two main resources of contemporary capitalism. To study and understand these roles, it is first helpful to draw on the theory of systems. Technological innovation cannot be understood when isolated from its context. Putting innovation in perspective requires a holistic and systemic approach: new technologies, new products, but also new markets, new organization and new management practices. The authors in this volume stress the importance of cognitive and social linkages, as well as the role of 'communities of practice' in the assessment and analysis of innovation within the enterprise, local and national economies and at the international level. The approach of scholars as well as practitioners should be directed toward the economic and social impacts of innovation activities.

The authors here perform a series of studies at different levels (research labs, enterprises and networks) and throughout systems of innovation and innovative milieux based upon practical cases, to present emerging ideas from established theoretical developments of economists, historians and geographers. Indeed, the genesis of knowledge and its diffusion (information) depends upon the density of the relationships among individuals and organizations at both micro-economic and macro-economic levels. What may be concluded from the analysis is that some specific actors (the state, enterprises) determine separately but also jointly, the process of diffusion, coordination and standardization of knowledge and technologies. Institutional and entrepreneurial networks play a fundamental role (vis-à-vis the employee or the entrepreneur) in the explanation of the systemic nature and the dynamics of innovation within contemporary capitalism. Cognitive relations between the actors of innovation show that the issue of appropriation of knowledge is as important as the production of new knowledge for the genesis of innovation.

The permanent innovation strategy of the enterprise needs continuous flows of scientific and technological inputs. Externalities and agglomeration effects are two phenomena associated with location that must be taken into account in the constitution and constant renewal of the enterprise's scientific and technological potential. Location choice is a precursor to the reinforcement of the firm's innovation capacities, allowing it to reach new markets and to penetrate complex networks (enterprises, research centres, other institutions). These networks constitute pools of resources for large enterprises and are seen as an essential tool for the creation of new innovative businesses and for regional development.

The growing link between knowledge and markets can be considered as symbolic of the current socialization of production forces. The systemic approach is today highly relevant because of this socialization. The need is for constantly renewing products, services and production methods, and this increases the cost and risk of knowledge development and dissemination. Networks (systemic relations) among an increasing number of institutions (firms of different size, public and private research labs) are thus becoming obligatory for those who want to remain competitive. Innovative milieux thus become one of the key features of present-day systems that are developed throughout the book, and most explicitly in Part III.

The second key notion that permeates this book is the role of power, both economic and political. The free market economy is based on abstract hypotheses that can be called into question by reality. Starting from the renewal of technologies and markets, the authors in this book clearly show that military programmes and industries constitute one of the engines of capitalist economies, arguably even more than the weight of big firms and the regulatory force of the state. In point of fact, the interactions between the military sector and competitive industries feed innovation systems (at the micro- and macro-economic levels) with new technologies and new organizational methods. Given the importance of military R&D in the large economies for generating radical innovations, a central role is assigned in Part II of this book to the analysis of the functioning of the systems of military and civil innovation.

However the role of power also exercises itself in more subtle ways, as all of the chapters that follow indicate. Chapter 1 by Stefan Kesting develops McCloskey's notion of the power of persuasion in communication, as independently developed by Boulding out of using communication for conflict resolution (through reconciliation, compromise, etc.), in parallel with the well-known work of Hirschman on 'voice'. As Kesting observes, 'Inspired by Galbraith's book, *The Anatomy of Power*, Boulding distinguishes basically three forms of power: (1) Political-military, threat, or destructive power; (2) Economic, exchange, or productive power; and (3) Social, love, or integrative power.' Boulding defines the third of these as 'the power to be accepted, respected, legitimated, loved, and to form part of a larger network'. In practice it can involve social exclusion as well as consensus and inclusion. The argument is clarified further through reference to Myrdal's notion of cumulative circular causation and Habermas' concept of communicative action, and the ways in which the latter transforms into communicative power, as well as Galbraith's 'conditioned power' and 'countervailing power'. These ideas are shown to underpin familiar concepts from the literature on innovation such as routines (Nelson and Winter), scripts (Nooteboom), leadership (Witt), 'ba' space (Nonaka and Konno), learning by interacting (Lundvall), etc.

Chapter 2, by Marie-Pierre Bès, has as its objective '. . . to demonstrate that knowledge being developed in laboratories is also the object of conflicts concerning its sharing, development and publication between public sector teams and their industrial partners.' This draws on the postulate that '. . . tacit knowledge acquired in scientific laboratories is not "visible" and identifiable except during activities involving strong interpersonal relationships'.

Research networks of the kind studied in this chapter fall into the category of 'emerging networks that are formed within projects where neither the partners nor the contents are stabilized – the knowledge is local, competing, appropriable, specific and non-substitutable' rather than those that 'are consolidated, dominated by routine activities and groups of stabilized members'. The author studies the issues surrounding the difficulty of appropriating knowledge (intellectual property rights, or IPR) in such contexts, and the arrangements resorted to in order to permit communication, with special reference to joint public–private laboratories in France. These do not limit themselves to internal interaction but have to remain flexible in relation to other networks, both national and international.

The issue of proprietary versus open-access knowledge is taken further in Chapter 3, by Marc Isabelle. The author's survey shows the conflicts in scientific communication arising out of a heightened emphasis on IPRs in public research organizations (PROs) anxious to patent their activities – conflicts between organizations and even within laboratories. About two-thirds of the respondents to the survey ('research directors' of a leading French PRO) are able to undertake both proprietary and open-access research.

Chapter 4, by Alfredo Ilardi and Blandine Laperche, examines the progress made towards an integrated patent system in Europe. Patents symbolize the well-known adage that 'knowledge is power'. Attempts to harmonize IPRs in Europe have been continuing for more than a century and are linked to patterns of globalization. Conflicts of interest have continually slowed such attempts, despite the resort to political action to overcome them. The chapter concludes by expressing the view that supportive political actions should be enhanced at the European level.

As already noted, Part II of the book (Chapters 5 to 7) deals overtly with military power, and since it is discussed at length below we pass on here to Part III, where communication and relational issues return via innovative milieux and clusters, in which the balance between networks and power relations becomes analytically more complex. Dimitri Uzunidis in Chapter 8 defines a network as 'a system whose purpose is to achieve integrated production (often multiple) in a single value chain and under the direction or coordination of holding or central firms'. Others however consider 'pure' networks as relationships – particularly for the interchange of knowledge – in which both market and hierarchy relationships are absent. In practice, such 'pure' networks are rare in the commercial and industrial world, where markets, hierarchies and networks tend to be interconnected and overlapping, in ways that have yet to be fully defined in the analytical literature. Uzunidis' chapter assesses the role of the individual enterprise in terms of the balance between investing in its local environment (the innovative milieu) and drawing upon it, as a potential arena for conflict and indeed conflict resolution.

For Schumpeter, 'The entrepreneur is prompted by a desire for power'. In Chapter 9, Sophie Boutillier, Blandine Laperche and Dimitri Uzunidis carry forward this idea into asking what constrains the emergence of entrepreneurial actions, and why the role does not endure. In their view the heroic entrepreneur of Schumpeter has become a 'socialized entrepreneur' in a context of networked capitalism. In this respect the entrepreneur has not really disappeared, but his/her power structures have altered considerably – knowledge plus finance and connections dictate the emergence of the power of the entrepreneur, but other and more pragmatic elements also become important in preserving that power.

Examples of such elements are evident in Chapter 10, by Kavoos Mohannak and Robyn Keast, on science parks, a format that in many parts of the world has failed to deliver its promises. In the case they study, of the Brisbane Technology Park in Australia, they found that the structures in place give rather too much emphasis to developing internalized cooperative networks, whereas they found more need to establish external linkages. A new balance may have to be struck between maintaining strong ties and facilitating weak ties (in the sense of Granovetter).

Finally, Chapter 11, by Paul Sommers, on the life sciences cluster in the Seattle (Washington state) area of the USA, assesses the role of power asymmetries in the 'cluster', in respect of the relationships between large pharmaceutical companies ('big pharma') and small biotechnology labs, universities, non-profit organizations and government and venture-capital funds in this region. Formal associations and partnerships (such as the 'Prosperity Partnership') have done a great deal to integrate activities and diverse agents that could have been little more than a local 'clump' into a properly constituted 'cluster'. The state's objective seemingly remains to establish new, powerful integrated corporate hierarchies, but whether or not this will happen is an open question.

PART I: CORE LINKAGES IN THE GENESIS OF INNOVATION: THE KNOWLEDGE DIMENSION

The first part of the book deals with the core linkages which allow the genesis of innovation. It focuses particularly upon the production, the diffusion and the appropriation of knowledge.

In Chapter 1, Stefan Kesting draws attention to the interaction among individuals that is at the origin of the development of knowledge and innovation. He links communicative action concepts (from Boulding, Habermas, Hirschman, Myrdal) with those of innovation studies (like Schumpeter, Nelson and Winter, Nooteboom, Penrose), in order to show the innovative capacities of communication and its role in the production of new knowledge. The innovative capacities of communication rely on three arguments: information is obtained in conversation, new ideas emerge in dialogues and debates, and institutional change is initiated and implemented on the basis of public disputes. Communication is at the core of the creation, processing and transmission of routines. The innovative capacities of teams are identified as a major resource of the firm. The innovation capacities of networks and clusters are also based on the interaction amongst their members. After an introduction to the model of communicative action, the discussion which follows paves the way towards a communicative theory of innovation.

Communication is particularly important as in today's knowledge-based economy, research relies on collaboration between different types of institutions (notably, from both the public and the private sector). This new context may be a source of advantage (sharing of experience and cost reduction) but it may also result in tensions dealing with the appropriation of shared knowledge. Chapter 2, by Marie-Pierre Bès, deals with cognitive externalities and their flows among research teams. The study is based on a detailed analysis of 133 research contracts signed by the engineering laboratories of the French National Centre of Scientific Research (CNRS) between 1960 and 2000. The author shows how teams can rely upon these different contracts to source a specific form of expertise. More precisely, the experimental knowledge available in the labs and in particular their capacity to design and carry out instrumentation experiments is essential although difficult to quantify. Today's types of partnerships such as joint laboratories or technological platforms are a good example of the attempts made to lock up externalities within restricted places.

The question of appropriability of research is also dealt with by Marc Isabelle, whose main objective in Chapter 3 is to provide a typology of research activities connected to the new context of the knowledge-based economy. Public research systems have experienced deep transformations worldwide since the 1980s. In European countries, they are now strongly challenged by the Lisbon strategy of the knowledge economy, a prominent engine of knowledge production and dissemination. Among the changes that are taking place, researchers are prompted to incorporate greater considerations for economic and societal needs into their activities, as well as better protecting their results and working in closer connection with private firms. Some scholars advocate this new social contract for science whereas others express concerns about the pitfalls that may arise from it. In his contribution of detailing 'Pasteur's Quadrant', Donald Stokes had pushed the debate forward by showing that basic research could be use-inspired – and actually was in many cases. Yet, Marc Isabelle argues that Stokes' taxonomy misses a dimension – namely that of proprietary vs. open-access research – which must be taken into account for fully grasping the economics of how the research sector may foster the knowledge economy. The expanded taxonomy allows better identification of the directions in which the pressure for increased relevance of public research really goes and to discriminate among the underlying motivations. The case study of the CEA (Commissariat à l'Energie Atomique) yielded preliminary results concerning this issue and demonstrated that changes have occurred in the way research is conducted, due to the specific nature of the linkage between knowledge and markets.

As networking becomes more pervasive in the course of the current socialization of production, so the need for protection of invention is also growing. Intellectual property rights are incentives and protection tools. In Chapter 4, Alfredo Ilardi and Blandine Laperche retrace the history and evolution of an integrated patent system in Europe and explain this development through the collective and global character of the innovation process in a knowledge-based economy. As the genesis of innovation depends on international – and regional – links between several actors, integrated protection systems are also needed. Ilardi and Laperche also describe the difficulties of harmonization: notably, in the case of the 'Community patent', technical difficulties and the political character of the protection of inventions. To obtain better results in terms of innovation, it is necessary, according to these authors, to link the reflection upon patent law harmonization with innovation policies and to transform innovation into a real European political priority.

The first part of the book shows that the new context of the knowledge-based economy has resulted in the tightening up of systemic links between research labs (in either public or private sectors) and the market. Information (communication) and knowledge (appropriation) are the two pillars of the innovation process that mobilize various actors. These new links have offered new opportunities for the production, sharing and dissemination of knowledge, and they also suggest constraints in the relationship to the appropriation of research results.

PART II: MILITARY-BASED INNOVATION NETWORKS

In Chapter 5, Claude Serfati addresses some issues related to military and commercial technology relations. He puts the debate into a historical context, and emphasizes how dialogues on these relationships have changed over the past six decades, going from 'spin-off' to 'spin-in', then to 'dual-technologies'. In the mid-1990s it was widely thought that the generalization of dual-use technologies would become a final note in the story of the relationship between military and commercial technologies. However, the Internet case study suggests a more complex relationship between military and commercial dimensions than the usual claim that the Internet 'was invented by the military'. Moreover, the emergence of security issues which are at the core of the globalization process provides a doctrinal framework against which the trajectories of new technologies are emerging, marking a new era in the history of the relationship between military and commercial technologies. It is against this background that, according to Serfati, the new security agenda adopted by the European Union (EU) and the USA must be analysed. This new security agenda combines, in a new and rather unexpected way, military and civilian technologies in some significant sectors (the space sector in the EU, and the biotechnology sector in the USA). Five criteria are relevant for investigating the relations between military and commercial technologies: history, international balance of power between countries, sectors and technologies, entrepreneurship, and contingency.

Within this new context emerges the concept of 'interface structures' that refer to a new type of organization to promote and increase interactions between the different elements of an innovation system, especially between research establishments and potential users of their capabilities. Theses interface structures are studied by Elena Castro-Martínez, Ignacio

Fernández-de-Lucio and Jordi Molas-Gallart in Chapter 6. The authors identify similarities in the strategies adopted by organizations involved in the transfer of knowledge in two different countries (the UK and Spain) and contexts (university–industry and military–civil sectors). Each is confronted with a common problem: the need to bridge dissimilar organizational cultures and practices. Without any contact among them, these organizations have developed remarkably similar and innovative strategies. Based on their experience, these authors define and characterize the 'interface structure' as a catalyst, increasing the interactions between different types of actors.

Due to the increase of complex linkages between the military and the civil sectors and technologies, it is interesting to study the impacts of different and changing institutional environments upon the firm's knowledge base. In Chapter 7, Michel Callois analyses the case of the defence industry as a specific institutional environment. Based on a sample of 600 observations collected among the best performing defence firms from 1999 to 2005 and their portfolio of patents obtained from the European Patent Office database, Callois argues that the institutional environment shapes the knowledge of defence firms. He proposes a methodological issue to highlight the influence of the institutional context upon patenting behaviour. The trade-off between civilian and military activities is expressed by the relative share of defence income in comparison with the global income of the firm. Econometric tests are run to assess these propositions. They show that the structure of the knowledge base differs depending on the institutional context. Callois concludes that there is a narrow co-evolution between the firm's behaviour and institutional environment. 'Pure' defence firms have a more targeted knowledge base (where capabilities are focused on a few activities) whereas 'mixed' or pure civilian firms have a broader knowledge base.

The second part of the book reveals that the production of military knowledge is more and more embedded in innovation networks that are not specific. This increases the complex character of innovation systems.

PART III: FROM KNOWLEDGE TO MARKET: SYSTEMIC LINKS AT THE LOCAL LEVEL

In Chapter 8, Dimitri Uzunidis studies the logic of the innovative milieu. For him, the innovative milieu is the socio-economic environment of a territory built by history and is the result of interactions among firms, institutions and human resources. These interactions stem from reciprocal synergies (networks, partnerships) between different local agents (from the public or the private sectors) specialized in economic and industrial development. An analysis starting from the innovative milieu provides an opportunity to study the enterprises' environment in order to understand their innovation dynamics. The systemic nature of relations which characterizes an economic and social milieu explains the incentives or the constraints to the act of innovation. Can innovation then be reduced, as a production of the milieu, to mere exchanges among individuals resulting in new productive combinations? Does innovation result from a specific organization of economic relations? Uzunidis' thesis is that the systemic nature of the milieu not only refers to economic interactions but also takes into consideration the social structures that are at the origin of innovative behaviours. In turn, the innovative milieu contributes to the firm's innovative performance, by supplying and organizing the flows of scientific and technical resources.

The contribution of Chapter 9, by Sophie Boutillier, Blandine Laperche and Dimitri Uzunidis, focuses on entrepreneurship, which can be considered to be a form of innovation. The question raised by these authors is the following: why does the entrepreneur's function, in the sense of J.A. Schumpeter, disappear? That is, why it is not long-lasting? Thanks to an approach linking economics and sociology, the construction of what the authors call his 'potential of resources' is revealed to be the set of knowledge, relations and financial resources that the entrepreneur builds in the context in which he or she acts. This potential of resources is itself not stable and may be increased or diminished, in various economic, political and social contexts. The success of entrepreneurship stems from the networks the entrepreneur creates. The formalization in terms of the 'organic square of entrepreneurship', proposed by these authors, explains the temporary character and territorial roots of the entrepreneur's function.

The next two chapters study two cases of innovative milieux and are especially focused, on the one hand, on the roles of exchange between the actors of a local system of innovation, and on the other, on the importance of the institutional structure which may boost or hinder the genesis of innovation. In Chapter 10, Kavoos Mohannak and Robyn Keast study the role of cooperative networks of high-tech SMEs within a technology park. By using a case study of an emerging South-East Queensland (Australia) technology park to provide insights, the chapter focuses on the nature of relationships and exchanges between SMEs and their stakeholders. Drawing on interviews and responses to questionnaires, including a network linkage survey and other documentary materials, this study explores the role of networks and clusters, and sheds some light on the degree of linkages between SMEs and other industry and research agents within the Brisbane Technology Park. The context and conceptual framework used for the analysis of networks within the technology park is centred on the relevance of two key concepts for SME development: 'innovation clusters' and 'cooperation networks', recognizing that both are emerging as significant tools to promote regional development through the activation, diffusion and expansion of locally generated knowledge.

In Chapter 11, Paul Sommers raises a significant issue: what are the differences between a clump (a mere aggregation of firms) and a cluster? Is the cluster idea a useful construct for thinking about regional development issues? To answer these questions, he uses a case study of biotechnology and life sciences in the Seattle area. The hypothesis stemming from the pattern of development of biotechnology in life sciences (rapid technological change, uncertain processes involving several institutions, importance of tacit knowledge, role of venture capital, etc.) is that clusters should be an efficient form of industry organization and that clumps should either be less successful or not present in the industry. Sommers explains public and private roles in cluster development around Seattle, notably in the creation of incentives to collaborate. Collaboration and exchanges of tacit knowledge are the key elements distinguishing a cluster from a clump, the latter being more based on competition. To conclude, Sommers presents different policy alternatives for a cluster based on research activities for the Seattle case. All (for example the development of biotechnology firms into fully integrated pharmaceutical companies, or the strengthening and the widening of the existing biotechnology cluster) require involvement by strong state and regional economic development organizations, in both short-term and long-term structural policies. To create new regional competitive advantages in the global context requires a fresh start based on existing local conditions, as well as insights about future development options for each region.

Innovation systems, whether local or national, are open systems. Their strength certainly depends on the density of their internal links, but the challenge they face also depends upon their ability to adjust to external evolutionary patterns (firms' strategic options, public policies, general economic climate). As a whole, then, the aim of this book is to reveal the role of the systemic approach to the genesis of innovation at the micro-economic level (inter-individual relations connecting knowledge production units) and at the meso-economic level that binds these together (relationships among institutions within sectors or local systems of innovation). Power relations and asymmetries, in both military and civilian spheres, have to be given full weight alongside the growing emphasis on networking and clusters. Specific lessons can then be drawn from this contribution of networks and particular innovation systems regarding global economic performance.

PART I

Core linkages in the genesis of innovation: the knowledge dimension

1. Towards a communicative theory of innovation

Stefan Kesting

1. INTRODUCTION

The classical conceptualization and definition of innovation in economics originated in the works of Joseph Aloïs Schumpeter. Schumpeter emphasized that truly welfare-enhancing economic development is not based on a smooth adjustment from a slightly disturbed equilibrium to another, but is based on revolutionary innovations (Schumpeter [1911] 1934, p. 64). He defined innovation as new combinations which lead to new products, new ways of production, discoveries of new resources, new organizational methods of running the business enterprise and entry into new markets (Schumpeter [1911] 1934, p. 66). In the process of innovation, path-breaking inventions and discoveries (technical, managerial or others) are picked up by radical individuals called entrepreneurs and turned into commercial successes. Schumpeter assumed the 'new' was somehow emerging out of the blue or floating around (ibid., p. 88). The distinctive role of the entrepreneur is radically turning away from traditions, customs and routines and committing herself/himself to push for change.

One of the few neoclassical economists who take discourse seriously, is McCloskey. She calculates that about a quarter of national income is produced by 'persuasive talk' (including, for example, advertising, see McCloskey and Klamer 1995, p. 194). In her book, *Knowledge and Persuasion in Economics*, she collects a bulk of evidence for the economic significance of persuasion under the heading 'The economy as a conversation'. McCloskey (1994, p. 370) uses the example of Donald Trump to point to the power of persuasion and 'the art of felicitous speech acts' to close deals. For McCloskey, this power of persuasion is the outstanding characteristic of Schumpeter's entrepreneur, for it is he or she who persuades banks to invest in innovations (ibid., p. 372).[1]

Schumpeter explicitly and analytically distinguishes between the entrepreneur and the investor or banker: 'He stands between those who wish to form new combinations and the possessors of productive means'

(Schumpeter [1911] 1934, p. 74). According to Schumpeter, whilst persuading the investor to lend him/her money, the entrepreneur also shifts the burden of uncertainty towards the provider of financial funds: 'It also settles the question whether the ordinary shareholder as such is an entrepreneur, and disposes of the conception of the entrepreneur as risk bearer' (ibid., p. 75).

Schumpeter offers a set of motives as an alternative, or more precisely supplement, to the neoclassical view (ibid., p. 93). This set of motives elucidates why the entrepreneur is more willing and able than others to take a radical turn. However, while constructing his theory of 'creative destruction' and accumulation (Schumpeter 1942, pp. 31, 32), he left two elements not fully explained:[2] the origin of the new and how the entrepreneur actually accomplishes the transformation of the new combination into a commercial success. At least concerning the first of these two tasks, Schumpeter did not develop a distinctive microeconomic theory of dynamic entrepreneurial activity.[3]

How have these conceptual gaps been addressed in the theory of innovation so far? How can such contributions be improved? The thesis put forward in this chapter is that the dynamic process of innovation leading to improvements in economic development cannot be fully understood without considering the basic human capability of communication. The discussion of the innovative capacities of communication will rely on three crucial distinctions:

- Information is obtained in conversation (the 'syntactic' aspect of communication).
- New ideas emerge in dialogues and debates (the 'semantic' aspect of communication).
- Institutional change is initiated and implemented on the basis of public disputes (the 'pragmatic' aspect of communication, Watzlawick et al. 1967).

2. BOULDING'S COMMUNICATIVE THEORY OF ECONOMIC ACTION

Kenneth E. Boulding developed a microeconomic theory based on communicative action which can be used as the basis for a communicative theory of innovation. The following sub-sections will assess in turn Boulding's three major theoretical building blocks of his conception of communicative economic action in an inclusive social process: 'image', 'conflict resolution', and 'integrative power'.

2.1 The Image

Boulding's theory of 'the image' contains an alternative microeconomic approach compared to the neoclassical one which is, apart from terminology and one crucial theoretical difference, quite similar to Denzau and North's mental models approach (1994; for a detailed comparison, see Kesting 2007). A short, and at the same time inclusive, definition of the image can be found in an article by Warren Samuels: 'The fundamental role of the image is to define the world. The image is the basic, final, fundamental, controlling element in all perception and thought. It largely governs our definition of reality, substantively and normatively, in part as to what is actual and what is possible' (Samuels 1997, p. 311). Hence, instead of rational maximization the image carried in our heads largely governs our behaviour in a semantic and pragmatic way (Watzlawick et al. 1967): 'The behavior is response to an image, not a response to a stimulus, and without the concept of an image the behavior cannot possibly be understood' (Boulding [1956] 1997, p. 43). All economic transactions are embedded in a process of 'imagination' which incorporates value judgements: 'The image of value is concerned with the *rating* of the various parts of our image of the world, according to some scale of betterness or worseness. We, all of us, possess one or more of these scales. It is what the economists call a welfare function' (ibid., p. 11).

It is left open whether the image remains stable or changes depending on the experiences of the individual and the influence of outside messages on it: 'The image is part of – and changes within – an ongoing process in which experience and image and selective perception coevolve' (Samuels 1997, p. 312). How does our image containing our value judgements, preferences and welfare perception change and how do others influence that process? Boulding insists that human beings communicate with each other via symbols (that is communication that can become independent of the communicator) as well as face-to-face (Boulding [1956] 1997, p. 88), and that communication constitutes and changes our images: 'It is this symbolic image and the communications which establish it and which change it which constitutes the peculiar quality of human society, a quality which no animal society shares' (ibid., p. 44). Samuels highlights the linguistic character of the image, too: 'Images are linguistic phenomena for mankind. Language is the material of images. Having an image reducing it to words, talking about it – all this involves the use of language' (Samuels 1997, p. 317). Image and language alike are at the same time intra- and inter-individual and their change takes place via correspondence of these levels: 'The basic bond of any society, culture, subculture or organization is a "public image" that is, an image the essential characteristics of which are shared by the individuals participating in the group. . . . Indeed, every public image begins in the mind of some single individual and only becomes public as it is transmitted and shared' (Boulding [1956] 1997, p. 64). It is not only true for the images of individuals, however, but also for shared images, 'that there is not a single public image, but there are many public images, as many indeed as there are cultures and subcultures within the great frame of the human race' (ibid., p. 132). In fact, a subculture is nothing else than a group of people sharing a certain public image. What has this micro-theory of behaviour guided by individual and public images which are stabilized or changed through communication to contribute to the development of an alternative approach to innovation?

What we regard as a 'new combination' in the Schumpeterian sense is a value-laden public image with the help of which we try to change and improve economic processes and their outcomes. Our image also helps us on the individual level to deal with uncertainty, which is done by the interplay of the image with habits and routines of behaviour:

> In the course of the repetition of habitual activities changes occur in the value structure, that is, in the value image of what we are doing. As we contemplate in good, rational manner the alternatives which are presumably open to us today the likelihood that we will select something else than the habitual pattern depends, of course, upon our satisfaction with this pattern. If we were miserable yesterday we are much more likely to assess the nonhabitual alternative favorably than if we were happy. At some point the misery in the contemplation of the habitual behavior overcomes the uncertainty involved in contemplating any other kind of behavior and we make a reorganization. (ibid., p. 87)[4]

On the collective level the novel creation is attained by a:

> process of the mutual modification of images both relational and evaluational in the course of mutual communication, discussion, and discourse. The course of the discussion is punctuated by decisions which are essentially temporary in nature in the sense that they do not close the discussion, although they do, of course, have the effect of modifying it. In one sense, in a successful political process all decisions are interim. We live in a perpetual state of unresolved conflict. A decision is partial resolution of conflict. It should never be a complete resolution. (ibid., p. 103)

According to Boulding the new is not only created, but also propagated and implemented in processes of communication which change public imagination. These processes are not free of conflict. The explanatory advantage of such an image- and communication-based theory of innovation is not only that it includes changing preferences[5] and interdependence of individuals, but also that it allows for welfare-enhancing effects through social learning.

2.2 Conflict Resolution

Boulding not only dealt with the impact of language and image on innovation, but also with the related topics of conflict resolution and power. Whereas the image represents the micro-foundation of his theory of 'new combinations', his view of conflict resolution can be interpreted as a complementary form of coordination. In a short paper entitled 'The Communication of Legitimacy', Boulding states on the one hand, the potential of communication or certain rituals like handshakes to put even parties in a hierarchical power relationship on an equal footing – the profound social truth, as he calls it, '. . . that communication can only take place among equals' (Boulding 1974, p. 240). On the other hand, he sees a Veblenian form of communication, where speech is used to symbolize a certain position or status in society. This power-oriented use of language might be labelled as 'conspicuous conversation', paraphrasing Veblen ([1899] 1967).

In his book, *A Preface to Grants Economics*, Boulding painted three pictures that clarify the role of communication in exchange and gift relationships (see Figures 2.1, 2.2 and 2.3 in Boulding 1981, pp. 20, 22, 23). The basic idea is simply that A sends B something and B sends something else back in return. Nevertheless, in addition to neoclassical economics the material and monetary flows (solid arrows) are accompanied or even 'paid for' by communications, information, threats, promises, affirmations, persuasions, and so on (dotted arrows) (ibid., p. 19). Boulding argues that: 'In communication, of course, the possibility that what leaves one party will not be what the other party receives – that is, that X_a will not be the same as X_b, and that Y_b will not be the same as Y_a – is very strong' (ibid., p. 21). Thus, apart from adversity of interests, simple misunderstandings may lead to conflicts. If we take into consideration that the images of different subcultures of a society are not the same, conflicts between them are also likely.

Because 'conflicts may be bitter and destructive, or they may be fruitful and constructive' (Boulding 1962, p. 306), Boulding looks for ways of resolution and institutions that would produce rather the latter kind of conflicts. He presents five basic ways of conflict resolution: avoidance, conquest, award, reconciliation and compromise. The first, 'avoidance', is almost exactly parallel to Hirschman's 'exit' option. Avoidance stands for the fact that if one party in a conflict leaves, the conflict will no longer exist. Equally as for Hirschman's exit interpretation, for Boulding markets are mechanisms of conflict resolution by the way of avoidance (ibid., p. 308). Conquest is just an extreme form of avoidance (respectively exit), where one party of the conflict is forcibly removed completely (ibid., p. 309).

If the parties can neither conquer nor avoid each other, a form of procedural resolution of conflict that is some form of 'voice' is likely. One widely used form of procedural conflict conclusion is jurisdiction. Boulding calls it 'award' and defines it as a settlement, where 'both parties have agreed to accept the verdict of an outside person or agency' (ibid., p. 310).

For Boulding the ideal type of procedural conflict conclusion would be 'reconciliation', 'in which the value systems of the images of the parties so change that they now have common preferences in their joint field: they both want the same state of affairs or position in the joint field, and so conflict is eliminated' (ibid., p. 310). This convergent modification of the images of the two or more parties is the result of conversation, argument, discussion, or debate. To reach reconciliation or consensus as one might also call it, according to Boulding certain conditions during discourse have to be met.

1. The value images of the two parties have to be not completely rigid (ibid., p. 311).
2. A success of the reconciliation process is more likely if the conflict occupies more the shell than the core of the value image of any party involved.
3. Instead of threatening the person of the other party, concern for the rival would ease reconciliation.
4. A culture where consensus is highly valued would obviously back up such a solution (ibid., p. 312).
5. This culture or experience may be reflected more in some personalities than in others (ibid., p. 313).
6. An independent third party (a conciliator or mediator) facilitates a consensus solution (ibid., p. 316).

If reconciliation is impossible, 'compromise' would be the second-best solution of a conflict. Compromise according to Boulding would be aimed at, if the value systems of the parties are too distant from each other to reach consensus, however each party is still willing to settle for something less than their ideal position. In that case mediation and conciliation may ease the process of bargaining and lead to compromise. Note that, though Boulding is distinguishing analytically between reconciliation and compromise, he nonetheless sees them amalgamated in practice: 'Frequently, however, both reconciliation and compromise go on together; indeed, some reconciliation may be necessary before compromise is possible' (ibid., p. 310).

Quite similar to Hirschman's description of the strength of the party that is left with voice as its only option (Hirschman 1970), Boulding mentions that the special commitment of the weakest bargainer gives him or her frequently the strongest bargaining position (Boulding 1962, p. 315).[6]

A third party or mediator might be necessary or useful to conduct the process toward a peaceful and mutually advantageous solution. The role of the mediator, as Boulding sees it, is to clear up misunderstandings in an intense emotional field. But it may not be enough for the neutral third party to facilitate communication between the opponents. The mediator should also introduce solutions that neither party has previously contemplated or envisaged (ibid., p. 318). Boulding stresses the prospect of institutionalizing mediation for the sake of the future of our societies and the necessity to follow certain rituals and procedures step by step during a process of conflict resolution. It takes time (that is transaction costs) to get to consensus or fair compromise.

Whereas Hirschman (1970) showed the potential of voice to invent and set in motion creative solutions for social problems, which would improve our societies, Boulding stresses the productivity of language for the resolution of conflicts. He takes a step further than Hirschman in distinguishing different forms of voice and considering that discourse is a procedural concept. Additionally Boulding's conditions for successful reconciliation or compromise contain a more detailed prescription of the attitudinal among other elements necessary to bring about the productivity of language for economic and social action. Last but not least Boulding brings in the mediator. The facilitating role of the neutral third party is not a part of Hirschman's voice concept.

2.3 Integrative Power – the Power of Persuasion

Boulding's approach can be seen as a microeconomic foundation for the Schumpeterian and Austrian process of innovation and discovery (Hayek 1969). Boulding is in general very much interested in a consensus-oriented development of society. As he writes in his book, *Three Faces of Power*: 'Economic development is fundamentally a learning process and learning is on the whole non-conflictual, though it has some conflictual elements in the elimination of error and a possible threat to personal identity which this may involve' (Boulding 1990, p. 196). Conflict is in most cases unproductive for him. For this reason he introduced the idea of conflict resolution in the USA. My interpretation focuses on Boulding's language-based integrative power, which is related to image and conflict resolution.

Inspired by Galbraith's book, *The Anatomy of Power*, Boulding distinguishes basically three forms of power: (1) Political-military, threat, or destructive power; (2) Economic, exchange, or productive power; and (3) Social, love, or integrative power. Boulding's power concept is very broad and goes even beyond social and economic analysis. Related to human behaviour, however, he defines the three power forms as follows: 'I distinguish between personal destructive power, which also involves the power of threat; personal productive power, which also involves the power to exchange; and personal integrative power, which is the power to be accepted, respected, legitimated, loved, and to form part of a larger network' (Boulding 1990, p. 79). Integrative power, which is related to image and conflict resolution, is defined by Boulding as follows: 'Integrative power depends very much on the power of language and communication, especially on the powers of persuasion' (ibid., p. 221). 'Integrative power often rests on the ability to create images of the future and to persuade other people that these are valid' (ibid., p. 122). Additionally he points out that integrative power 'is the most dominant and significant form of power' (ibid., p. 10, and see ibid., p. 110) compared to the other two forms.

Though integrative power rests on the use of language, speech as its source does not – as Habermas ([1981] 1995) would argue – create and inherently include a consensus-oriented form of rationality, but it is linked to and enhanced by a variety of emotional relations, which according to Boulding can be ordered on a scale that reflects their intensity: reciprocity, respect for each other, the wish and willingness to learn from each other, sympathy for each other, and love for each other. Integrative power, however, must not always be consensus-oriented or inclusive. Boulding writes about several ways in which it can be used during a conflict against others:

1. In a network or group that was built up by integrative power, some persons may achieve a powerful position or status;
2. 'Language can be a powerful weapon of destruction in putting people down, in complaining, in nagging, in recriminating' (Boulding 1990, p. 81);
3. It can be used to stigmatize and exclude people: 'The power of social exclusion is a very important aspect of the overall integrative system' (ibid., p. 85);
4. The ability to persuade people can be used to manipulate them: 'Unfortunately, what is convincing is not always true, and what is true is not always convincing' (ibid., p. 119).

Nevertheless integrative power is a consensus-oriented concept, which bears a lot of similarities with Habermas's communicative power (Habermas [1992] 1999).

2.4 Conclusion

What is the relation between Boulding's three theoretical elements of image, conflict resolution and integrative power; and how do they combine into an alternative approach to understand processes of innovation?

To create new combinations they first have to be imagined, often while communicating in groups or teams (Penrose 1959). These images may nonetheless differ from individual to individual. They are, however, also connected via integrative power which is based on emotional bonds and communication: 'Integrative power within such a network may be a gradual process of a person gathering respect from the other people in the network through communication and the formation of images of that person in the minds of the other network members' (Boulding 1990, p. 114).Via integrative power (networks) inter-subjective public images are formed and changed. Because it is very likely that individual images and public images of subcultures are in conflict with each other, some form of conflict resolution (mediation) is necessary to facilitate and legitimate this process and organize it in a peaceful way. A necessary condition of such an interactive process is, however, that the participants have an open attitude towards each other: 'Perhaps the most important single source of integrative power could be described as the capacity to love in a generalized sense, which means a capacity not only to be aware both of the broader and the narrower environment around a person but also to find these environments attractive and interesting, and to put a substantial value on them, especially in terms of benevolence. This means . . . that the person perceives an increase in his or her own welfare when he or she perceives an increase in the welfare in some sense in the surrounding world' (ibid., p. 115). Hence, innovation is potentially more successful if communicated as mutually beneficial. A figure which integrates the aforementioned elements of Boulding's conception would look like Figure 1.1.

3. MYRDAL'S VIRTUOUS AND VICIOUS CUMULATIVE CIRCULAR CAUSATION

The way Gunnar Myrdal conceptualizes 'democratic planning' (Myrdal 1968) is very similar to Boulding's formation of public images through integrative power. He explains his discursive conception of human action in the appendix, 'A Methodological Note on Valuations and Beliefs' in his book, *The American Dilemma*. Myrdal starts from a concept of an individual whose actions are determined by his or her value judgements. These judgements may not only be in conflict from person to person, but also intrapersonally: 'In this treatise, therefore, behavior is conceived of as being typically the outcome of a moral compromise of heterogeneous valuations,

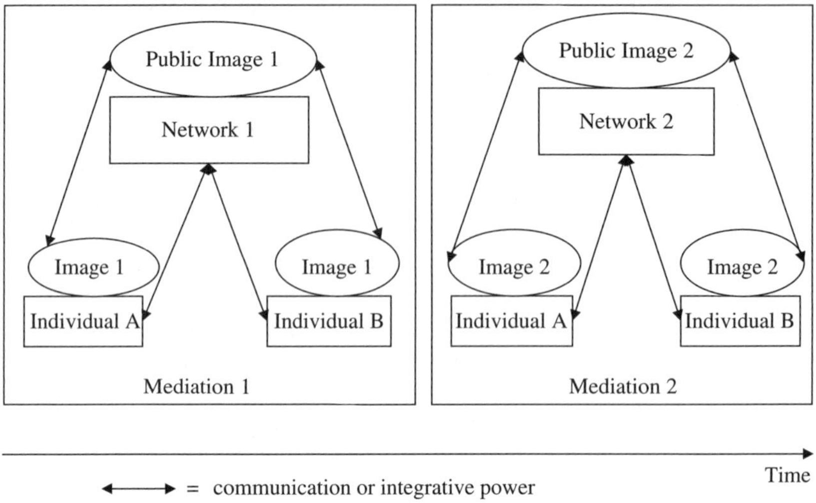

Figure 1.1 Boulding's concept of change

operating on various planes of generality and rising in varying degrees and at different occasions to the level of consciousness' (Myrdal [1944] 1972, p. 1028).

People do not become aware of intra- and inter-personal conflicts coincidentally; instead it happens while communicating with others: 'The individual or the group whose behaviour we are studying, moreover, does not act in moral isolation. He is not left alone to manage his rationalizations as he pleases, without interference from outside. His valuations will, instead, be questioned and disputed. Democracy is a "government by discussion", and so, in fact, are other forms of government, though to a lesser degree' (ibid., p. 1028).

This exchange and critique of each others' arguments as an attempt to persuade and convince takes place on a background of a life-world which is at least partly shared by participants in discourse: 'When discussion takes the form of moral criticism by one person or group of another, it is not that the one claims to have certain valuations that the other does not have. It is rather an appeal to valuations which the other keeps in the shadow of inattention, but which are assumed, nevertheless, to be actually held in common' (ibid., p. 1028).

> By this democratic process of open discussion there is started a tendency which constantly forces a larger and larger part of the valuation sphere into conscious attention. . . . A need will be felt by the person or group, whose inconsistencies

in valuations are publicly exposed, to find a means of reconciling the inconsistencies. This can be accomplished by adjusting one of the conflicting pairs of valuations. (ibid., p. 1029)

This conception sounds familiar to someone acquainted with Habermasian discourse ethics. The urge to clarify inconsistencies through adjustment and modification is what Habermas sees as the forceless force of the better argument (compare also Biesecker 1997; Ulrich 1997; Habermas [1981] 1995). Though Myrdal is aware of a tendency to neglect and deny the existence of value conflicts,[7] he is relying on the fascination of new ideas, which will bring about a trajectory towards more rationality: 'The direction in a normal and peaceful process of popular education is toward decreasing inconsistency' (Myrdal [1944] 1972, p. 1033).[8]

To sum up, Myrdal stresses the importance of participation in democratic discourse processes in solving value conflicts and directing economic development.

As we have seen, scholars working in the tradition of Old or American Institutionalism developed a range of notions of communicative processes and their role in innovation and institutional change. Boulding calls these gains from communicative cooperation 'public images': 'The basic bond of any society, culture, subculture or organization is a "public image that is, an image the essential characteristics of which are shared by the individuals participating in the group. . . . Indeed, every public image" begins in the mind of some single individual and only becomes public as it is transmitted and shared' (Boulding [1956] 1997, p. 64). Myrdal is also stressing the transaction benefits of clarifying one's interests and preferences.

Contemporary American Institutionalist authors who elaborate the transaction benefits of clarifying and solving conflicts of interests and values include Knight (1992, 2001); Knight and Ensminger (1998); Samuels et al. (1997); and Schmid (2004).

4. THEORY OF COMMUNICATIVE ACTION: HABERMAS'S DISCOURSE ETHICS

The specific innovative productivity of speech that brings about a specific kind of rationality and thus allows the creation of mutual benefits during a transaction can be clarified further by drawing on Habermas's Theory of Communicative Action. Though Habermas is not an economist and his behavioural pattern of communicative action is not directed to the economic realm, his theory can nonetheless improve Hirschman's, Boulding's and Myrdal's attitudinal and procedural concepts of communicative action.

Communicative action in Habermas's sense is not only oriented to success, efficiency, or personal goals, but also to reaching an understanding among the participants of a discourse and is coordinated 'through cooperative achievements of understanding among participants. In communicative action, participants are not oriented primarily to their own success, but to the realization of an agreement which is the condition under which all participants in the interaction may pursue their own plans' (Habermas [1981] 1995, p. 541; see Biesecker 1997, p. 220). Communicative action is based on language and operates in the process of discussion. This procedural exchange of arguments during which participants learn to understand each other's motivations, underlying norms, and opinions is called 'discourse' by Habermas. In discourse, participants are required to learn from each other and to change their own attitudes toward the world in general or toward certain problems occurring within it.[9]

So far, Habermas's concept of communicative action or discourse is similar to the procedural conflict resolution of reconciliation described by Boulding. What Habermas adds is a certain procedural communicative rationality[10] that helps to differentiate three basic types of arguments (speech acts) which can be criticized or defended, grounded in their specific rationality. Habermas argues that communicative rationality occurs inevitably during discourse, which is evident if we thoroughly consider the inter-subjective meaning of illocutions. If we try to persuade during discourse, we suppose that the other person can be convinced by our arguments and will accordingly change his or her mind. When we do this, however, we implicitly concede that exactly the same might happen to us but in the opposite direction. That is to say, we would admit the superiority of the other's arguments and change our minds.[11] The communicative rationality of speech acts is not only instrumental, like the utilitarian rationality of economic man, but threefold. As Adelheid Biesecker put it (quoting and translating Habermas), speech acts 'are not simply grounded in knowledge of the object world (as in empirical thinking), but also in the norms of the society in which the discourse is taking place (Habermas's social world) and the values of the partners in the discourse (Habermas's subject world). Communicative rationality, therefore, has three dimensions: An action [or a statement, S.K.] is rational if it is objectively true, socially right and subjectively sincere' (Habermas [1981] 1995, p. 149; translated in Biesecker 1997, p. 220).

The participants of a discourse use their shared experiences (made in their life-world) as background and reservoir to test the validity of arguments along the three dimensions of rationality just mentioned. In a certain discourse situation, the discussants refer to their shared experiences, which contain all opinions and world views taken for granted to begin a cooperative process of interpretation and new combination. During this process, some elements of their experiences will remain untouched or stable, while others will become a matter of doubt and may change. Because discourse, as a form of social coordination, is linked to the social and subjective worlds, it has the capacity to integrate other values than those occurring during a neoclassical or Austrian market process and transaction, which is merely based on instrumental rationality and where values have to be reflected monetarily in supply or demand to influence the outcome.

For the development of creative solutions of conflicts or problems discussed in a discourse, following communicative rationality, Habermas and Habermasian economists[12] would rely on all the persons participating in discourse. This establishes the special innovative productivity of the discursive process.

The philosophical underpinning of the argument is that participants in a dialogue logically have to allow for mutual understanding to occur or they will commit a 'performative contradiction'. German discourse ethic philosophers Apel and Habermas argue that, if you want to influence through persuasion (that is exchange of arguments), you assume that this person will change his or her mind after hearing what you have to say. If you start from this anthropological vantage point you must believe that this will hold for you conversely. That is, you are also willing to change your mind if convinced by the other person's arguments. Hence, to assume opportunism or strategic behaviour in dialogue is what Apel and Habermas would call a 'performative contradiction'. What is true for persuasion in one direction of influence will in turn also be true for learning in the other direction. If you try to receive information (useful ideology or mental models for instance[13]) from a person, you assume that people are capable of learning. Conversely you should assume that the other will also be able to learn from whatever information you do provide in the process of communication.

Philosophically, a strong Kantian, a priori emphasis on reason is apparent in discourse ethic. This bright-side view of discourse is probably as one-sided as the dark-side view of interpreting it as solely strategic (see Kesting 1998, 2005). Empirically both opportunistic behaviour (manipulation, betrayal, false and broken promises and so on) and mutually rewarding co-operation are possible and observable in actual transactions and processes of innovation. They are a complex mix of exploration and exploitation (Nooteboom 2000a, p. 304).

5. TRANSFORMATION OF THE DIALOGICAL MODE INTO COMMUNICATIVE POWER

Habermas is thinking along a rather strict theoretical dichotomy of two modes of communication, termed 'arguing' and 'bargaining' by Jon Elster (1998), who thinks in a similar kind of fashion. On one side there is the exchange of rational arguments to deliberate about ends and means while in this process transforming the preferences of participants. On the reverse side there is bargaining, where the outcome of the process results from the bargaining power of the parties – that is, the resources that enable them to make credible threats and promises.

In his book, *Between Facts and Norms* ([1992] 1999), however, Habermas shows awareness of the fact that both modes are intertwined in concrete discourses in the public sphere. It is not possible do justice to Habermas's extensive treatment and explanation of the complex philosophical and sociological relation between communicative action in the public sphere and the constitutional, legal and parliamentary political system of Western democracies in that book. The purpose is merely to highlight the way in which he conceptualizes the transformation of communicative action into communicative power. The public sphere and deliberative politics within it is a problem-solving debate and a conflict of interests at the same time. Even actors who have little strategic power (to bribe, buy, or have easy access to media or impose a threat) can turn communicative action into communicative power, that is a countervailing power against opposing strategically powerful actors (Habermas [1992] 1999, p. 381). In discussing Arendt's power theory Habermas writes: 'But discursively produced and intersubjectively shared beliefs have, at the same time, a *motivating* force. Even if this remains limited to the weakly motivating force of good reasons, from this perspective, the public use of communicative freedom also appears as a generator of power potentials' (Habermas [1992] 1999, p. 147). Moreover, after his discussion of Elster's arguing and bargaining distinction he concludes: 'The results of deliberative politics can be understood as communicatively generated power that competes, on the one hand, with the social power of actors with credible threats and, on the other hand, with the administrative power of officeholders' (Habermas [1992] 1999, p. 341). A good chance for communicative power to have an impact on the course of social development appears especially in moments of crisis. Habermas explains:

> But, in general, one can say that even in more or less power-ridden public spheres, the power relations shift as soon as the perception of relevant social problems evokes a *crisis consciousness* at the periphery. If actors from civil

> society then join together, formulate the relevant issue, and promote it in the public sphere, their efforts can be successful, because the endogenous mobilization of the public sphere activates an otherwise latent dependency built into the internal structure of every public sphere, a dependency also present in the normative self-understanding of the mass media: the players in the arena owe their influence to the approval of those in the gallery. (ibid., p. 382).

Hence, there is an arena for communicative innovative processes in the public sphere where powers of persuasion are used.

Galbraith's perspective on power shows a remarkable resemblance with Habermas's. However, he stresses that it is neither positive nor negative. In his book, *Anatomy of Power*, he explains: 'Power can be socially malign; it is also socially essential' (Galbraith 1983, p. 13). The author distinguishes three forms of power. First, condign power which proceeds from threat, second, compensatory power which is based upon reward, and finally conditioned power which 'in contrast, is exercised by changing belief. Persuasion, education, or the social commitment to what seems natural, proper, or right causes the individual to submit to the will of another or of others' (ibid., pp. 5, 6).

Of course conditioned power is the interesting one for the question addressed in this chapter, because condign and compensatory power is nothing other than strategic action, whereas Galbraith's third category describes a language-based form of power. Galbraith identifies each form of power with a certain source of power (personality, property, and organization) and elaborates the complex relations between sources and forms, with considerable overlap and interdependence. However, he states boldly that modern industrial societies are in the age of organization and therefore conditioned power is the dominating form of power. Conditioned power can be understood as communicative action in Habermas's sense, when a group or organization uses discourse to reach a common understanding, and later publicly defends this consensus. One might say the members of an organization develop a persuasion to allow them to fight for that opinion much more powerfully than any person could do alone. Thus, conditioned power is consensus-oriented as well as conflict-oriented. It describes the transition from Habermas's communicative action to communicative power.

Galbraith writes that indirect conditioned power (traditions, patriarchy) are a subconscious influence, as part of the framework of values that surround and influence all members of society. Hence, Galbraith's conditioned power is used in overt, covert and latent conflicts (Lukes 1974). Dugger's theory of corporate hegemony, for instance, is built upon Galbraithian thought[14] and can be interpreted as conditioned power in the extreme.

Interesting, and a little confusing at first, is Galbraith's conclusion from his analysis of the workings of conditioned power: Galbraith writes: 'As we have sufficiently seen, organization and the associated role of social conditioning are basic to all modern exercise of power. At the same time, and paradoxically, they bring not only the modern concentration of power but also its *personal* diffusion' (Galbraith 1983, p. 183).

The diffusion of power that Galbraith mentions results from his concept of 'countervailing power' and his persuasion that conditioned power is often not in fact power (as an influence on someone), but only an illusion of power. Galbraith explains: 'There are few manifestations of power in modern times that expend such costly and committed energy as the cultivation of belief and the resulting exercise of power through advertising. However, partly because advertising is a wholly ostentatious attempt to capture belief, it is not a fully reputable way of winning it. It regularly invites its own resistance and disapproval' (ibid., p. 30).

In other words, attempts at 'creative destruction' may be completely futile. They might meet considerable resistance because disapproval may also result from countervailing direct conditioned power issued by consumer activists or environmentalists. The case of Brent Spar (Post et al. 2002) shows the general openness to direct conditioned power and the related openness of public discourse concerning its results and succeeding decisions.

Hence, from Galbraith's and Habermas's theories we gather that dialogical practice can generate or strengthen communicative power. In other words, if it is used successfully in a public debate, it will change the minds of people participating in it and thereby also change the range of ideas, norms and ideologies which constitutes a process of innovation (see Figure 1.2).

6. COMMUNICATION IN THE THEORY OF INNOVATION: COMMUNICATED ROUTINES AND SCRIPTS

Such a theory of communicative innovation as developed above is already apparent as an undercurrent in Richard Nelson and Sidney Winter's concept of routines. In their 1982 book, *An Evolutionary Theory of Economic Change*, they describe organizational routines as analogues of individual skills and as an important part of a company's memory and therefore its knowledge. They stress the importance of routines for the smooth operation of any organization and highlight their tacit knowledge component. Skills and routines allow for an automatic and habitual response, and hence for fast and efficient handling of all kinds of challenges and tasks (ibid., p. 97). On the level of skills, when reasoning about the choice for a particular task and on

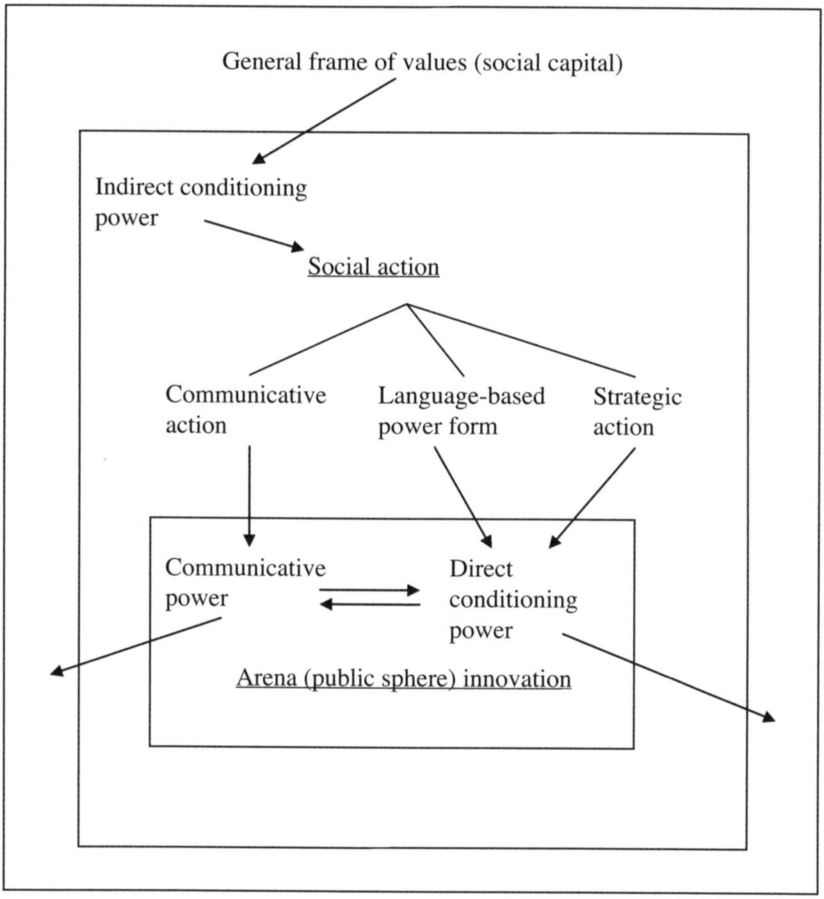

Figure 1.2 Power as forms of social action

how to change and learn skills, Nelson and Winter point to the importance of language: 'It is, as we have emphasized, difficult or impossible to use language to characterize the "inner workings" of a skill, but words serve quite well in thinking and communicating about skills considered as units of purposive behaviour' (ibid., p. 85). Moreover, in considering problem-solving skills as learning they write: 'In the exercise of these cognitive skills, an important role is played by language and, in particular, by the names of other skills that may or may not be possessed by the planner or problem solver' (ibid., p. 86). According to them, uncertainty about the right word that is the right skill for the circumstances might lead to *semantic* ambiguity (ibid., p. 88).

On the level of the organization, appropriate routines are coordinated and 'triggered' by a constant stream of signals and messages, hence based on language: 'For organization members in such roles, there are additional requisites of knowing the job that parallels the ones involved in receiving and interpreting such messages. These include, again, the abilities to speak and write the natural language of the society to which the organization belongs, but also the important additional requirement of command of the organizational dialect' (ibid., p. 102). Especially because of their tacit knowledge component and because they constitute socially shared knowledge, routines as part of the organizational memory are founded on explicit and implicit forms of communication: 'To view organizational memory as reducible to individual member memories is to overlook, or undervalue, the linking of those individual memories by shared experiences in the past, experiences that have established the extremely detailed and specific communication system that underlies routine performance' (ibid., p. 105). Surprisingly, Nelson and Winter do not discuss the role of communication and use of ordinary language when it comes to control, replication and imitation of routines. In a typical Schumpeterian way they point out that: 'Innovations in organizational routine similarly consist, in large part, of new combinations of existing routines' (ibid., p. 130); however, in their discussion of heuristics and search routines, they do not consider debate, argument and discussion to pool knowledge, to reach workable conclusions and to find appropriate solutions. Such a 'linguistic turn' is equally absent when Nelson and Winter 'review the behavioural foundations of the evolutionary approach' (Nelson and Winter 2002, p. 25) in an article tracing twenty years of theoretical development since their book was published.

However, this next analytical step is taken by Bart Nooteboom who, especially in one of his articles (Nooteboom 1999) and his book, *Learning and Innovation in Organizations and Economies* (2000a), replaces Nelson and Winter's 'routine' by the equivalent but analytically improved concept of a 'script': 'The actions of people, in turn, based on their cognition are substituted into nodes of organizational scripts, where the nodes represent tasks and organizational scripts represent "routines" or "performance programs"' (ibid., pp. 126, 127).

Nooteboom illustrates the concept of the script using the example of dining according to two different scripts: 'In a restaurant script, the nodes represent successive activities of entering, sitting down, ordering, eating, paying and leaving. Each can be done in different ways [substitutions]. For example, paying by cash, by cheque, bank card, credit card, or chip card' (ibid., p. 126). An alternative script is the following: 'Take for example, a self-service restaurant as opposed to a service restaurant. The sequence of

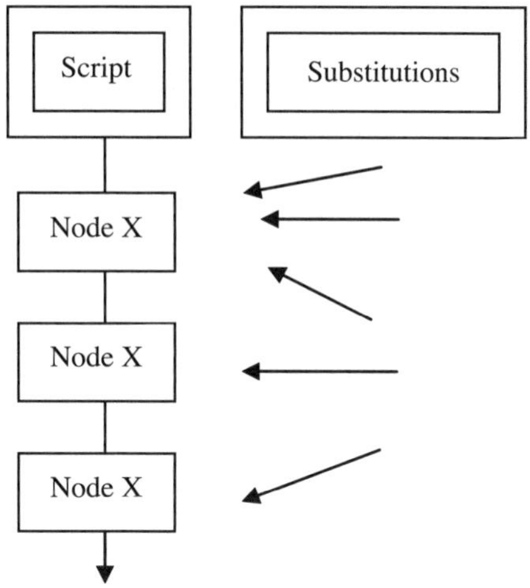

Source: Figure 6.1 in Nooteboom 2000a: 126

Figure 1.3 Scripts

nodes is different: in contrast with the service restaurant, one generally pays before sitting down and eating' (ibid., p. 128).

Founded on a trans-disciplinary discussion of modern cognitive science, Nooteboom argues that such scripts have to be seen as developed and applied in processes of 'interactive emergence' which are innovative discourses in the Habermasian sense:

> Knowledge and meaning constitute repertoires from which we select combinations in specific contexts, which yield novel combinations that may shift repertoires of knowledge and meaning. Such shifts of knowledge and meaning occur in interaction with the physical world, in technical tinkering, and in social interaction, on the basis of discourse (cf. Habermas's [1981] 1995, 1984, notion of 'communicative action'). (Nooteboom 2000a, p. 121)

Nooteboom explicitly subscribes to 'methodological interactionism' (see Figure 3.3 of his book), which is based on communicative action: 'In fact, the book aims to replace both methodological individualism of economics and the methodological collectivism of (some) sociology with methodological interactionism' (ibid., p. 303).

Nooteboom explains how innovation in scripts and routines occurs through the change of meaning in metaphor and by the hermeneutic circle. According to him – and quite similar to Boulding's concept of the image – using a metaphor allows us to take a different perspective or point of view:

'Things that seemed the same are now seen as different, and things that seemed different are now seen as the same. To effect such a break is, I propose, the function of metaphor, which carries over a feature from one word to another and thereby sheds a different light, which might cause us to form a novel concept by redistribution of features' (ibid., p. 140). 'Metaphor also plays an important role in reducing cognitive distance, in making something intelligible from one cognitive framework to another, by expressing something from one framework in terms of the other' (ibid., p. 145). However, in this process of understanding and new combination we are enabled and constrained at the same time by the use of language as can be illustrated in the hermeneutic circle:

> Gadamer (1977) is the recognized 'father' of hermeneutics. The basic notion, according to him, is that, like an institution more generally, language provides an 'enabling constraint' (my term not Gadamer's). It enables us to interpret and understand but thereby also constitutes a prejudice, and interpretation that we construct and impose rather than an objective meaning that we find. (ibid., p. 142)

To sum up, Nooteboom's groundbreaking and sophisticated approach provides us with further elements to build a foundation for a communicative microeconomic theory of innovation.

7. TEAMS, NETWORKS AND CLUSTERS

This micro-foundation of communicative action implicitly and explicitly underlies discussion and analysis in several theoretical variants of research on innovation. It can be found in concepts of leadership (Witt 1998), teams (Penrose 1959; Nonaka and Konno 1998), communities (Cohendet and Diani 2006), clusters (Porter 1998), organizational, regional as well as global networks (Burt 2004; Kogut 2000; Ruef 2002; Granovetter 2005; Castilla et al. 2000; Saxenian 2000, 2003; and Thompson 2003) and last but not least, innovation systems (Lundvall 2002).

Some of these attempts to explain innovation as an interactive process emphasize the dialogical mode of communication, where the exchange of arguments fulfils the creative purpose of combining knowledge to arrive at new ideas and solutions for problems, others highlight the more strategic use of speech acts to convince and persuade others of discoveries and initiate and push for institutional and organizational change. Some just focus on one of these aspects of communicative action or rely implicitly on such a micro foundation.

In his concept of leadership, Witt combines both modes of communicative action. He stresses 'the fact that individual learning is socially conditioned so that tacit cognitive commonalities emerge within intensely communicating groups' (Witt 1998, p. 68). However, such a creative process is given direction by visionary entrepreneurs: 'the institutional set-up of the interactions assigns certain individuals a position in which they get a chance to shape the communicative process and thus to exert an influence on the collective outcome' (ibid., p. 71). This according to Witt constitutes leadership, which has to strike a careful balance to allow enough freedom for innovation as well as imposing control to channel it in the desired direction.

At the core of Penrose's notion of the role of teams to overcome uncertainty and to discover and exploit opportunities for a firm lies a notion of communicative action. Only via interaction to combine local knowledge and expertise can a team become more than just the sum of its parts (Penrose 1959). To integrate new employees and transform them into valuable team members is costly (ibid., p. 45).

The communicative nature of innovation in teams is made explicit by Nonaka and Konno in their description of 'Ba' – the method of innovation employed by Japanese firms. According to them team building starts with the integration of new members in a process of socialization (Nonaka and Konno 1998, p. 42). On the basis of this first stage a dynamic of knowledge sharing, extension and creation can start which they call 'externalization' and 'combination'. Both are based on a concept of communicative action. 'Externalization requires the expression of tacit knowledge and its translation into comprehensible forms that can be understood by others' (ibid., p. 43). This leads to a fusion of an individual's ideas and intentions into a shared mental world or group image: 'the articulation of tacit knowledge – that is, the conversion of tacit into explicit knowledge – involves techniques that help to express one's ideas or images as words, concepts, figurative language (such as metaphors, analogies, or narratives), and visuals. Dialogue, "listening and contributing to the benefit of all participants", strongly supports externalization' (ibid., p. 44). The next step is equally dependent on communication: 'Combination involves the conversion of explicit knowledge into more complex sets of explicit knowledge. In this stage, the key issues are communication and diffusion processes and systemization of knowledge' (ibid., p. 44). Hence, the dialogical mode of communication is the microeconomic foundation which explains the creative and innovative advantage of teams.

Lundvall distinctly points to the interactive nature of innovation in his account of the Danish innovation system: 'Rather, innovation comes about as the result of a social process involving an interplay between many individuals and organizations over a longer period of time in which cumulative learning processes take place' (Lundvall 2002, p. 43; see also Nooteboom 2000b). For these processes to be effective, they have to be embedded in 'social capital – the ability to work with and trust others' (ibid., p. 99). As Nooteboom demonstrated in his book on trust, this invaluable social resource is built on dialogue and conversation (Nooteboom 2002, p. 195). An example of building trust via public deliberation is referred to by Lundvall:

> The Danish Council of Technology and the consensus conferences it organizes have a more central position in Denmark as a forum for instigating debate than corresponding institutions in other countries. Consensus conferences bring together panels of ordinary citizens with experts and the outcome is well reported in the press and used as input for parliament when it designs new regulations and laws. (Lundvall 2002, p. 26)

Clusters as conceptualized by Porter (1998) can be seen as regional and industry-specific sub-systems of innovation. Here trust (ibid., pp. 80–81) and dialogue (ibid., p. 89) are not their foundation, but rather a consequence of clusters based on geographical proximity. However, such an explanation based on mere geographical closeness lacks behavioural foundation.

Granovetter's work on networks seems to suffer from a similar lack of micro-foundation. He also stresses a causality running from social proximity (for instance: 'network density', 'strength of weak ties') toward 'shared ideas' and 'more novel information flows' (Granovetter 2005, p. 34) and not vice versa. At least the emphasis in his theoretical argument is on the enabling effect of ties for communication flows and not that ties (whether strong or weak) are built on communication (whether persuasive or dialogical).

Saxenian's rationale for the innovative success of networks is in part grounded in geographical proximity, which is, however, supported by an implicit concept of communicative action: 'The trust, information exchange and teamwork which are the basis of collaborative supplier relations require continued interaction which is difficult to achieve over long distances' (Saxenian 2000, p. 319). Note that she stresses continued interaction as well as closeness in this quote. The latter aspect moves into the background for obvious reasons in her more recent work on transnational entrepreneurs. They travel a lot to overcome distance, however the emphasis is on communication: 'Today's "Argonauts" depend upon travel and electronic connections that give real-time communications around the globe and allow the transfer abroad of skills, technology and business models faster and more flexibly than most large corporations can achieve' (Saxenian 2003, p. 97).

In his introductory chapter before discussing and comparing a variety of theoretical accounts for the workings of networks, Thompson particularly highlights verbal interaction: 'This reinforces the point that where it was a lack of communication between parties, this can easily threaten to undermine the fragile level of cooperation established. Cooperative networks thrive on communication and information flows between their members' (Thompson 2003, p. 115).

People who are able to facilitate such information flows in networks because they either connect problems with their solutions, or solutions with their implementation, or recombine bits of knowledge to create new ideas, are brokers bridging structural holes in Burt's understanding of networks. Their business is: 'information arbitrage' (Burt 2004, p. 354), to solve conflicts because they 'translate the belief or practice' (ibid., p. 355), 'to draw analogies' (ibid., p. 355) and 'synthesis' because they 'see new beliefs or behaviours' (ibid., p. 355). As Burt points out: 'brokers are critical to learning and creativity' (ibid., p. 356). Their service relies on either dialogue or persuasion.

In their analysis of 'knowledge-intensive communities' Cohendet and Diani (2006, p. 239) try to tackle the question what it is in their communication that leads to the success of communities in building trust, ties and new knowledge. Their definition of community seems to be relevant and captures all the aforementioned terms (leadership, teams, innovation systems, clusters and networks): 'In this contribution we will be considering genuinely "autonomous" communities based upon the principle of voluntary adhesion, due to the sharing of a number of values, norms or common interests' (ibid., p. 246). The crucial theoretical contribution in their article is a distinction between two fundamental variables to measure communication: 'the "degree of repetition of interaction" . . . and the "nature of communication"' (ibid., p. 252). The former depicts how often people talk to each other (frequency) while the latter accounts for the richness of the texture that is the content, of their communication (intensity). A low level of both leads to a 'weak communicative culture' while high frequency and intensity results in a 'strong communicative culture'. If empirically reliable measures for both frequency and intensity can be found this might allow studies in the future to predict and forecast the innovative success of communities, depending on the nature of their communicative culture.

8. CONCLUSION: A COMMUNICATIVE MICRO-FOUNDATION OF INNOVATION

As we have seen quite a number of economists refer sometimes explicitly, sometimes implicitly, to communication when they analyse particular aspects of innovation. Moreover, they usually start from some descriptive notion of verbal, symbolic or media-based human interaction without providing a clear definition of what communication actually is. To gain an even deeper understanding of the micro processes of discovering and constructing the 'new', however, it is important to start from a well-defined idea of communication as such, as Nooteboom does.

Communication between individual actors might be seen as potentially successful or as potentially unsuccessful. One prominent theoretical approach tends to stress that communicative action may lead to mutual understanding of a situation and hence, productive consensus (Habermas [1981] 1995; Biesecker 1997). Moreover, other scholars highlight that for innovation, conflict can be seen as useful, inspiring and productive as reaching a common understanding (Boulding 1990; Galbraith 1983).

Other widely used approaches emphasize that discourse is manipulative (Foucault 1983), or highlight – based on a notion of bounded rationality and an assumption of opportunisic behaviour amongst actors – that all communication is costly (Williamson 1991; Denzau and North 1994).

Demonstrating that for analytic purposes a more levelled approach proves to be more fruitful, some scholars try to combine the optimist and pessimist perspective on communicative action (Elster 1998; Flyvbjerg 1998).

What all these concepts of communication have in common is that they put particular forms of human relations at the centre stage of their analysis of the process of persuasion and not the question of how communication as such allows for the discovery or construction of the *new*. However, some scholars point out that a theory of communication which stresses the human relationship in interaction might not be the most promising approach to understand the role of communication in innovation as such.

The German sociologist Dirk Baecker is one of those critics. He proposes an alternative, cybernetic approach to communication. Instead of focusing on the action-oriented aspects of communication (that is rules, norms, motives and causal relationships) he proposes to view communication as specific to situations which are characterized by surprise (Baecker 2005, p. 8). In his theory Baecker wants to abstract from interaction of concrete and present human beings (ibid., p. 49). A student of Niklas Luhmann, he combines systems theory with cybernetics (Bateson 1972; Brown 1972; Foerster 1996; Luhmann 1984, 1990, 1995; Watzlawick et al. 1967; Wiener 1948, 1991) and proposes to built on Claude E. Shannon's mathematical theory of communication (Shannon 1948), which has been used so far primarily for information technology and for encoding and decoding information.

According to Baecker, Shannon's theory describes a recursive process of information processing. By assigning probabilities to incoming messages, it calculates the condition of the world in which one lives and at the same time conversely evaluates incoming messages according to a probabilistic anticipation based on the condition of the surrounding world. It separates a complex contingency into an indefinite sequence of single contingencies which consequently can be analysed (Baecker 2005, p. 19).[15] Linking this idea to Brown's *Laws of Form* (1972), Baecker arrives at his own social definition of communication, as a procedure of information processing which allows for an orientation in a contingent world. The order of this world can neither be assumed nor does it have to be questioned altogether. It can, however, be explored with the help of a sequence of a given set of contingencies. Based on von Foerster's cybernetics (1996) he argues that the participant observer in a constant stream of communication oscillates between the undefined and the definable, partly controlled, partly controlling the process.

It remains to be seen and is open to further enquiry whether such a cybernetic and systems theoretic approach can further advance our understanding of communicative innovation.

NOTES

1. This entrepreneurial skill even found its way into television, with shows like the *Dragon's Den* (BBC 2006).
2. Though he developed some ideas concerning this he did not publish them in his lifetime (Schumpeter 2005).
3. Nooteboom highlights this shortcoming of Austrian Economics in general: 'it deserts us when it comes to the clinch of understanding how knowledge shifts in interaction between "different minds thinking different things", because the black boxes of the minds of individuals and the knowledge of firms remain closed' (2000a, pp. 73–74), and: 'But note that Schumpeter also was unable to endogenize creation in the form of invention (cf. Witt 1993): innovation was the realization of potential offered by invention' (*ibid*., p. 63).
4. Note that this introspective process of evaluation explicitly includes an emotional component. An alteration of the habit is based on either feeling happy or miserable about it!
5. For a case study of an innovation involving preference change, see Wilfred Dolfsma's *Institutional Economics and the Formation of Preferences: The Advent of Pop Music* (2004).
6. A good example for this 'David-beats-Goliath' principle is the conflict around the Brent Spar. In 1995, the oil corporation Shell announced they would send their outworn oil rig Brent Spar to the bottom of the Atlantic Ocean and faced tremendous opposition organized by Greenpeace. At the end of that public conflict, Shell had to give in and scrapped the oil rig onshore (Post et al. 2002).
7. An observation, not only true for public discourse, but also for certain strands of economic theory, like for example the New Institutional Economics (Nooteboom 2000a, p. 112).
8. An apparent intellectual undercurrent in Myrdal's, Schumpeter's as well as Habermas's thinking is Max Weber's theory of rationalization [1921] 1978. For a critique of this view, see Zelizer (1997).
9. The open attitude toward arguments of the other, which is implied here as a precondition was called 'democratic ethos' by one of Habermas's critics: 'The time has come for

Habermas to acknowledge what pragmatists – especially Dewey and Mead – emphasized long ago: there is no democracy – in *theory* or in *practice – without a democratic ethos*' (Bernstein 1996, p. 1146).

10. To develop this type of rationality, Habermas draws on George Herbert Mead and two generations of ordinary language philosophy including the Speech Act Theory of John L. Austin and John R. Searle.
11. The discourse concept developed by Bill Waller and Linda Robertson shows some similarities to Habermas's (Waller and Robertson 1991, p. 1035).
12. I am indebted to Biesecker and Ulrich, who developed an economic theory based on Habermas's work, which they termed Practical Social Economics (Biesecker 1997; Ulrich 1987, 1997).
13. See Denzau and North (1994) and my critical interpretation in Kesting (2007).
14. Compare for example Dugger (1989) and his critique of Galbraith's work in Dugger (2001).
15. Baecker illustrates this method with a quote by Woody Allen: 'There is no question that there is an unseen world [a complex contingency, DB]. The problem is, how far is it from midtown and how late is it open? And after death is it still possible to take showers? [sequence of single contingencies, DB]' (Baecker 2005, p. 19, quoted from Allen 1997, p. 15).

REFERENCES

Allen, W. (1997), 'Examining psychic phenomena', in W. Allen (ed.), *Complete Prose*, London: Picador, pp. 13–22.

Baecker, D. (2005), *Form und Formen der Kommunikation*, Frankfurt am Main: Suhrkamp.

Bateson, G. (1972), *Steps to an Ecology of Mind: Collected Essays in Anthropology, Psychiatry, Evolution and Epistemology*, San Francisco, CA: Chandler.

BBC (British Broadcasting Corporation) (2006), 'Dragon's Den', http://www.bbc.co.uk/dragonsden/, accessed, 15 November.

Bernstein, R.J. (1996), 'The retrieval of the democratic ethos', *Cardozo Law Review*, **17** (4–5), 1127–46.

Biesecker, A. (1997), 'The market as an instituted realm of action', *Journal of Socio-Economics*, **26** (3), 215–41.

Boulding, K.E. (1962), *Conflict and Defence: A General Theory*, Lanham, MD: University Press of America.

Boulding, K.E. (1974), 'The communication of legitimacy', in K.E. Boulding (ed.), *Collected Papers: Toward a General Social Science*, vol. IV, Boulder, CO: Colorado Associated University Press, pp. 239–43.

Boulding, K.E. (1981), *A Preface to Grants Economics: The Economy of Love and Fear*, New York: Praeger.

Boulding, K.E. (1990), *Three Faces of Power*, Newbury Park, CA: Sage.

Boulding, K.E. (1997), *The Image: Knowledge in Life and Society*, reprint of 1956 edition, Ann Arbor, MI: University of Michigan Press.

Brown, S. (1972), *Laws of Form*, New York: Julian.

Burt, R.S. (2004), 'Structural holes and good ideas', *American Journal of Sociology*, **110** (2), 349–99.

Castilla, E., H. Hwang, H., E. Granovetter and M. Granovetter (2000), 'Social networks in Silicon Valley', in C.M. Lee, W.F. Miller, M. Gong Hancock and H.S. Rowen (eds), *The Silicon Valley Edge*, Stanford, CA: Stanford University Press, pp. 218–47.

Cohendet, P. and M. Diani (2006), 'Informal communication, collective beliefs and corporate culture: a conception of the organization as a "community of communities"', in R. Arena and A. Fertre (eds), *Knowledge, Beliefs and Economics*, Cheltenham, UK and Northampton, MA, USA: Edward Elgar pp. 239–66.

Denzau, A.T. and D.C. North (1994), 'Shared mental models: ideologies and institutions', *Kyklos*, **47** (1), 3–31.

Dolfsma, W. (2004), *Institutional Economics and the Formation of Preferences: The Advent of Pop Music*, Cheltenham, UK and Northampton, MA, USA: Edward Elgar.

Dugger, W.M. (1989), *Corporate Hegemony*, New York: Greenwood.

Dugger, W.M. (2001), 'Progress denied – the unravelling of the new industrial state', in M. Keaney (ed.), *Economist with a Public Purpose: Essays in Honour of John Kenneth Galbraith*, London and New York: Routledge, pp. 183–97.

Elster, J. (ed.) (1998), *Deliberative Democracy*, Cambridge: Cambridge University Press.

Flyvbjerg, B. (1998), 'Habermas and Foucault: thinkers for civil society?', *British Journal of Sociology*, **49** (2), 210–33.

Foerster, H. von (1996), *Wissen und Gewissen: Versuch einer Bruecke*, Frankfurt am Main: Suhrkamp.

Foss, N.J. (2000), 'Theory of the firm: contractual and competence perspectives' in N.J. Foss (ed.), *The Theory of the Firm: Critical Perspectives on Business and Management*, London and New York: Routledge, vol. IV, chapter 49, pp. 195–216.

Foucault, M. (1983), 'The subject and power', in H.L. Dreyfus and P. Rabinow (eds), *Michel Foucault: Beyond Structuralism and Hermeneutics*, Chicago: University of Chicago Press, pp. 208–26.

Gadamer, H.G (1977), *Philosophical Hermeneutics*, D.E. Linge (ed.), Berkeley, CA: University of California Press.

Galbraith, J.K. (1983), *The Anatomy of Power*, Boston, MA: Houghton Mifflin.

Granovetter, M. (2005), 'The impact of social structure on economic outcomes', *Journal of Economic Perspectives*, **19** (1), 33–55.

Habermas, J. (1984), *Vorstudien und Ergänzungen zur Theorie des kommunikativen Handelns*, Frankfurt am Main: Suhrkamp.

Habermas, J. (1995), *Theorie des kommunikativen Handelns*, paperbaek reprint of 1981 1st edn, Frankfurt am Main: Suhrkamp.

Habermas, J. (1999), *Between Facts and Norms*, translated by William Rehg, originally published 1992, Frankfurt am Main: Suhrkamp, Cambridge, MA: MIT Press.

Hayek, F. von (1969), 'Der Wettbewerb als Entdeckungsverfahren', in F. von Hayek (ed.) (1969), *Freiburger Studien, gesammelte Aufsätze*, Tübingen: J.C.B. Mohr.

Hirschman, A.O. (1970), *Exit, Voice, and Loyalty*, Cambridge, MA: Harvard University Press.

Kesting, S. (1998), 'A potential for understanding and the interference of power: discourse as an economic mechanism of coordination', *Journal of Economic Issues*, **32** (4), 1053–78.

Kesting, S. (2005), 'Countervailing, conditioned and contingent: the power theory of John Kenneth Galbraith', *Journal of Post-Keynesian Economics*, **28** (1), 3–23.

Kesting, S. (2007), 'Communicative Transaction Benefits', in W.R. Garside (ed.) *Institutions and Market Economies: The Political Economy of Growth and Development*, Basingstoke and New York: Palgrave Macmillan, pp. 322–42.

Knight, J. (1992), *Institutional and Social Conflict*, Cambridge: Cambridge University Press.

Knight, J. (2001), 'A pragmatist approach to the proper scope of government', *Journal of Institutional and Theoretical Economics*, **157** (1–2), 28–48.

Knight, J. and J. Ensminger (1998), 'Conflict over changing social norms: bargaining, ideology, and enforcement', in M.C. Brinton and V. Nee (eds), *The New Institutionalism in Sociology*, Stanford, CA: Stanford University Press, pp. 105–26.

Kogut, B. (2000), 'The network as knowledge: generative rules and the emergence of structure', *Strategic Management Journal*, **21** (3), 405–25.

Luhmann, N. (1984), *Soziale Systeme, Vol. I: Grundriss einer allgemeinen Theorie*, Frankfurt am Main: Suhrkamp.

Luhmann, N. (1990), 'Was ist Kommunikation?', in N. Luhmann (ed.), *Soziale Aufklaerung 6: Die Soziologie und der Mensch*, Opladen: Westdeutscher Verlag, pp. 113–24.

Luhmann, N. (1995), *Die Wissenschaft der Gesellschaft*, Frankfurt am Main: Suhrkamp.

Lukes, S. (1974), *Power – A Radical View*, Basingstoke and London: Macmillan.

Lundvall, B.Å. (2002), *Innovation, Growth and Social Cohesion – The Danish Model*, Cheltenham, UK and Northampton, MA, USA: Edward Elgar.

McCloskey, D.N. (1994), *Knowledge and Persuasion in Economics*, Cambridge: Cambridge University Press.

McCloskey, D.N. and A. Klamer (1995), 'One quarter of GDP is persuasion', *American Economic Review (Papers and Proceedings)*, **85** (2), 191–5.

Myrdal, G. (1968), *Beyond the Welfare State: Planning and its International Implications*, reprint of first edition, 1958, New Haven, CT and London: Yale University Press.

Myrdal, G. (1972), *An American Dilemma: The Negro Problem and Modern Democracy*, reprint of first edition, 1944, Harper and Row, New York: Pantheon Books.

Nelson, R.R. and S.G. Winter (1982), *An Evolutionary Theory of Economic Change*, Cambridge, MA: Belknap Press.

Nelson, R.R. and S.G. Winter (2002), 'Evolutionary theorizing in economics', *Journal of Economic Perspectives*, **16** (2), 23–46.

Nonaka, I. and N. Konno (1998), 'The concept of "Ba": building a foundation for knowledge creation', *California Management Review*, **40** (3), 40–54.

Nooteboom, B. (1999), 'Innovation, learning and industrial organisation', *Cambridge Journal of Economics*, **23**, 127–50.

Nooteboom, B. (2000a), *Learning and Innovation in Organizations and Economies*, Oxford and New York: Oxford University Press.

Nooteboom, B. (2000b), 'Institutions and forms of co-ordination in innovation systems', *Organization Studies*, **21** (5), 915–39.

Nooteboom, B. (2002), *Trust: Forms, Foundations, Functions, Failures and Figures*, Cheltenham, UK and Northampton, MA, USA: Edward Elgar.

Penrose, E.T. (1959), *The Theory of the Growth of the Firm*, Oxford: Oxford University Press.

Porter, M. (1998), 'Clusters and the new economics of competition', *Harvard Business Review* (November–December), 77–90.

Post, J.E., L.E. Preston and S. Sachs (2002), *Redefining the Corporation – Stakeholder Management and Organizational Wealth*, Stanford, CA: Stanford University Press.

Ruef, M. (2002), 'Strong ties, weak ties and islands: structural and cultural predictors of organizational innovation', *Industrial and Corporate Change*, **11** (3), 427–49.

Samuels, W.J. (1997), 'Kenneth Boulding's *The Image* and contemporary discourse analysis', in W.J. Samuels, S.G. Medema and A.A. Schmid (eds), *The Economy as a Process of Valuation*, Cheltenham, UK and Lyme, USA: Edward Elgar, 299–327.

Samuels, W.J., S.G. Medema and A.A. Schmid (1997), *The Economy as a Process of Valuation*, Cheltenham, UK and Lyme, USA: Edward Elgar.

Saxenian, A. (2000), 'The origins and dynamics of production networks in Silicon Valley', in R. Swedberg (ed.), *Entrepreneurship: The Social Science View*, Oxford and New York: Oxford University Press, pp. 308–31.

Saxenian, A. (2003), 'Transnational technical communities and regional growth in the periphery', in J.-P. Touffut (ed.), *Institutions, Innovation and Growth*, Cheltenham, UK and Northampton, MA, USA: Edward Elgar, 97–115.

Schmid, A.A. (2004), *Conflict and Cooperation – Institutional and Behavioral Economics*, Oxford: Blackwell.

Schumpeter, J.A. (1934), *The Theory of Economic Development*, originally published in German in 1911 in Leipzig by Duncker and Humblot, Cambridge, MA: Harvard University Press.

Schumpeter, J.A. (1942), *Capitalism, Socialism and Democracy*, New York: Harper and Row.

Schumpeter, J.A. (2005), 'Development', translated from an unpublished manuscript in German by Marcus C. Becker and Thorbjørn Knudsen *Journal of Economic Literature*, **43** (1), 108–20.

Shannon, C.E. (1948), 'A mathematical theory of communication', *Bell System Technical Journal*, **27**, 379–423 and 623–56.

Siebeck Paul (1978), 'Competition as a discovery process', in F. von Hayek (ed.), *New Studies in Philosophy, Politics, Economics and the History of Ideas*, Chicago, IL: University of Chicago Press, pp. 249–65 179–190.

Thompson, G. (2003), *Between Hierarchies and Markets*, Oxford and New York: Oxford University Press.

Ulrich, P. (1987), *Transformation der ökonomischen Vernunft: Fortschrittsperspektiven der modernen Industriegesellschaft*, Bern: Paul Haupt.

Ulrich, P. (1997), *Intergrative Wirtschaftsethik: Grundlagen einer lebensdienlichen Ökonomie*, Bern: Paul Haupt.

Veblen, T. (1967), *The Theory of the Leisure Class*, reprint of the first edition, 1899, New York: Macmillan, New York: Viking Penguin.

Waller, W.T. and L.R. Robertson (1991), 'Valuation as discourse and process: or, how we got out of a methodological quagmire on our way to purposeful institutional analysis', *Journal of Economic Issues*, **25** (4), 1029–48.

Watzlawick, P., J.H. Beavin and D.D. Jackson (1967), *Pragmatics of Human Communication*, New York: W.W. Norton.

Weber, M. (1921), *Economy and Society*, reprinted and translated 1978 by G. Roth. and C. Wittich (eds), Berkeley, CA: University of California Press.

Wiener, N. (1948), *Cybernetics, or Control and Communication in the Animal and the Machine*, Cambridge, MA: MIT Press.

Wiener, N. (1991), *Constructing a Social Science for Postwar America: The Cybernetics Group*, Cambridge, MA: MIT Press.

Williamson, O.E. (1991), 'Comparative economic organization: the analysis of discrete structural alternatives', *Administrative Science Quarterly*, **36**, 269–96.

Witt, U. (1993), 'Turning Austrian Economics into an Evolutionary Theory', in B.J. Caldwell and S. Boehm (eds), *Austrian Economics: Tensions and New Directions*, Boston, MA and Dordrecht: Kluwer, pp. 215–43.

Witt, U. (1998), 'Imagination and leadership: the neglected dimension of an evolutionary theory of the firm', *Journal of Economic Behavior & Organization*, **35**, 161–77.

Zelizer, V.A. (1997), *The Social Meaning of Money*, Princeton, NJ: Princeton University Press.

2. The division of scientific labour and the sharing of knowledge

Marie-Pierre Bès

In the 18th century, Adam Smith, upon witnessing the rise of the Industrial Revolution, explained that the division of industrial labour increased the global productivity of companies and thus led to the 'Wealth of Nations' through the integrated mechanization of production processes and its corollary, the specialization of tasks. Smith identified the savings of time from locating production on one site and from not having to move between sites as among the productivity gains from this specialization.

At the beginning of the 21st century, investments in research and development and in intangible goods (training, software, patents, and so on) have been identified by political leaders and organizations (the OECD, for example) as well as endogenous growth theorists (Romer 1986) as one of the major sources of economic growth. It is now widely recognized that the performance of a national economy is the result of dense relationships between public and private investments. Of major significance within the 'knowledge society' perspective of an innovation system are the links between industry and science. This refers to such different types of interactions as startups, collaborative research, contract research, development of intellectual property rights.

Nevertheless, the problem of the sharing of knowledge is mainly treated from the point of view of the balanced sharing of research results between 'open science' carried out mainly in public institutions (national research centres, universities) and 'closed science' carried out above all in the business sector (David 2003; Cassier 2002; Foray 2004) and as yet few articles on the scientific economy deal with the division of scientific labour.

The objective of this chapter is to demonstrate that knowledge being developed in laboratories is also the object of conflicts concerning its sharing, development and publication between public sector teams and their industrial partners. The study is based on a series of studies on industrial projects of the CNRS (Centre National de la Recherche Scientifique – French National Centre for Scientific Research) laboratories in areas of engineering sciences (Grossetti and Bès 2001; Bès 2004).[1] The two methodological postulates of this chapter are that tacit knowledge acquired in scientific laboratories is not 'visible' and identifiable except during activities involving strong interpersonal relationships (for example, training of doctoral students, contacts with industry, exchanges of researchers between organizations, creation of joint research labs) and from a longitudinal approach of individual careers and research collectives. Just as for routines in firms (Nelson and Winter 1982), it is on the site of the research itself (laboratories, companies, service providers) and through the analysis of the trajectories of the teams that we can understand the importance of this tacit knowledge.

In the first section we will present the nature of the networks and the cognitive resources activated through collaborative research: this involves explaining how 'applied' scientific knowledge involves a tacit, collective and instrumented know-how. Then we will show the potentialities or obstacles for the transfer, publication or appropriation of researchers' knowledge by the business sector. In fact, the last section of this chapter is devoted to joint CNRS/business laboratories viewed as systems for keeping the tacit knowledge of researchers in specific places, or sorts of 'enclosures' (Dumont and Holmes 2002).

1. NETWORKS AND COGNITIVE RESOURCES IN SHARED RESEARCH

Callon (1997) proposes that we consider two sorts of socio-technical networks: emerging networks that are formed within projects where neither the partners nor the contents are stabilized – the knowledge is local, competing, appropriable, specific and non-substitutable – and when networks are consolidated, dominated by routine activities and groups of stabilized members. Research projects fall into the first category, made up of open and unstable organizational configurations within which emerging knowledge and know-how circulate.

1.1 Individual Social Networks

Through the analysis of around one hundred joint research contracts established between CNRS laboratories and firms during the period from 1960 to 2000, we (Grossetti and Bès 2001) have shown the important role of social relations (44 per cent) in making contacts with partners, compared to the intervention of institutions (18 per cent) or using the science market (38 per cent). In 44 per cent of cases, the partners called on networks of personal relations (former colleagues, former students, friends, neighbours, family ties, and so on) to establish contact between a research team and a company. This can be explained by the formation of individual networks during studies or periods of professional mobility that are reactivated in the context of research and development. It is therefore plausible to explain direct links between the market for research and development contracts and the markets for training and employment. The three markets can be regarded as partially 'embedded'. We can therefore claim that collaborative research activities are based, in the initial contact phase, partly on social networks.

Moreover, when we study the length of these contracts, we cannot help but notice the length of the partnerships: 30 per cent last more than nine years and 4 per cent more than 23 years, even though it appears that the length of contracts has a tendency to decline over the studied period. A closer examination of the length of partnerships shows that there are non-contractual periods during which exchanges between partners continue to take place. These periods are characterized exclusively by individual relations and not by relations between organizations: these relations are qualified as 'informal' by economists. They confirm the importance of the phenomena of loyalty and mutual confidence between partners in the coordination of contracts (Joly and Mangematin 1996).

1.2 Cognitive Resources

Using data on the collaborative research activities of 23 CNRS teams, we have isolated four cognitive components that forge the expertise of a team (Bès 2005b):

1. Knowledge linked to a specific technology for a product or a process: for example, a team knows how to make and operate a customized micro-camera that can study certain properties of industrial products or processes and therefore facilitate quality surveillance operations.
2. Knowledge linked to the mastery of technical instruments: for example, a laboratory has unique equipment capable of measuring electrical voltage stress phenomena useful for testing the resistance of materials compliant with the technical specifications of the buyer.
3. Knowledge linked with scientific methodology (modelling, for example): a team knows how to model the movements of a robot within a given space and can therefore offer industrialists in the space sector programs adapted to moving automated machines.
4. Economic knowledge of one or more markets: for example, a laboratory that works with several companies in the food industry is familiar with common hygiene problems and can identify various market segments for its systems for purifying foods.

These four types of knowledge and know-how developed by research laboratories working with industrialists are idiosyncratic because they are highly instrumented (through experiments, manipulations, simulations), based on an original complementarity between researchers and technicians (mobile temporary workers and experienced permanent staff), and articulated around other research projects carried out by scientific leaders such as unsubsidized theses (Bès 2004) or longer projects. This is collective and tacit know-how as defined by Polanyi (1958) that is not entirely expressible or articulable. Certain know-how can temporarily take on visible explicit forms such as patents, new procedures or new companies, and other knowledge can be partially transmitted through the mobility of academic researchers to companies. Thus, the recruitment of PhD students can serve as a vector for the transfer of technology (Mangematin and Robin 2003). Other forms are highly tacit; that is, collective and articulated around specific material resources such as the design and carrying out of scientific experiments with the aim of improving industrial processes or the manipulation of sophisticated measuring devices (Latour 1987).

There is therefore cognitive apprenticeship on the part of teams through their industrial contracts that takes place essentially through the accumulation of competencies and the creation of a unique expertise around specific scientific equipment and tools. This apprenticeship is based on externalities of knowledge. Public scientific teams, which are more stable and lasting than their industrial equivalents, can strategically use several series of contracts and partnerships to create a niche in a scientific market (Grossetti and Bès 2001) and notably to manage their purchases of equipment effectively. In the same vein, the stability of public sector jobs is an advantage compared to the mobility of researchers in the private sector, which may hamper the formation of stable research groups. Hence, the research report published by the Centre d'Etudes et de l'Emploi (Research Centre on Employment) on the careers of researchers in private companies (Duhautois and Maublanc 2005) reveals the depreciation of the professional branch of scientific expertise. This is not the case for French industrial research for the period 1960–90 which together with public laboratories and state-run organizations formed a sort of National Innovation System (Mustar and Larédo 2002).

In research contracts signed with industrialists, the different components of knowledge deployed by researchers are only directly exploitable by companies that have a significant research activity (large groups, state-run technological organizations for example); but some requests on the part of companies, especially small and medium-sized businesses, concern only the use of this knowledge for targeted scientific studies applied to their industrial problems (Bès 2005b) rather than mastery of the knowledge itself.

1.3 Means of Appropriation and Exchange

A certain number of legal and contractual conditions concerning the ownership of the results, materials, and equipment (Cassier 2002), like the date of disclosure, the regular submission of research reports, or timetables, regulate the sharing of the partners' outcomes. But the knowledge acquired by the participants in a completed research project overflows beyond the publication of articles or patents.

It is accepted, in the literature relating to scientific activities (Bozeman 2000; Bozeman and Mangematin 2004), that the activities and scientific results of collaborative research are difficult to appropriate in their entirety by the organizations that have financed them (public or industrial institutions). But it is not only a problem of agency linked to the asymmetry of information between industrialists and laboratories,[2] there is also the presence of externalities of knowledge that can be found implicitly in the references commonly made to the knowledge or expertise of a research team.

As predicted by the economic theory of contracts, economic agents solve this secondary problem by setting up some original institutional arrangements that coordinate their exchanges. They cannot, in externalized situations, use classic contractual means. Neither the integration of the scientific team into the company nor the renewal of research contracts are business practices that guarantee the ownership or the global transfer of the 'goods' exchanged. In the first, rare case – the hiring of an entire team – the value of the scientific expertise also depends on the existence of the team within a particular environment (as part of an engineering school, a university, etc.) and in broader scientific communities. In the second case, the appropriation of scientific studies requires on the part of the firm the maintenance of a strong R&D capacity in order to absorb the results and the hiring of PhD students who contribute to laboratory competencies (Mangematin and Robin 2003).

We can also witness the development of two types of economic arrangements, one at the personnel level and the other at the organizational level:

- The temporary transfer of researchers between organizations: assignment of public researchers to companies, or of industrial researchers to public laboratories, co-financing of theses, employment of trainees in their final year of studies, generalization of postdoctoral research financed by industrial contracts.
- Temporary and flexible organizational structures such as R&D agreements or high-tech consortia (Cassier and Foray 2002) specifically created for some R&D projects. Powell (1990) has qualified these hybrid organizations as 'networks' in reference to sociological theories

on the embedding of economic relations within individual relations. Networks are forms of exchange characterized by the complementarity of the partners and the perpetual accommodation between participants: it is not only a question of respecting the terms of the contract but also of regularly adapting to the progress and new constraints of the participants. These original organizational forms, crisscrossed by individual relations, are like meta-networks that exist between the organizations.

2. THE CASE OF JOINT LABORATORIES WITH WALLS: ENCLOSURE OF KNOWLEDGE?

For a certain number of years in France, mixed research structures, also called joint laboratories, were created according to the American model which associates public and private laboratories for short periods (four years, renewable) in a common research project within a context of relative stagnation of investments in industrial and public research. The joint laboratories of the CNRS and industry provide interesting examples for the analysis of the tensions surrounding the appropriation of knowledge of industrial and public research teams.

Are these structures just temporary resilient forms (organizations linked together by 'invisible' relations woven through numerous contracts and strong individual relations) or an effort to reintroduce in private companies, activities that were previously subcontracted to public laboratories?

2.1 Governance of Joint Research Laboratories

On 31 December 2003 there were 25 joint laboratories between the CNRS and private companies in France, that is research contracts that for a limited time (in general four years) associated laboratories affiliated with universities with the CNRS and industrialists around a scientific programme, a specific budget and often a single site for experiments. According to the data provided by the CNRS, it is possible to underline the typical profile of a joint laboratory:[3]

- these organizations are recent creations (since 1990);
- three-quarters have been renewed;
- 70 per cent involve a large group;
- life sciences and chemistry represent 18 out of the 25 organizations;
- the majority are situated on a university campus or in laboratory buildings;

- in almost every case, CNRS joint research units are involved and very seldom joint service units.

A joint laboratory has scientific objectives that are finalized to meet the needs of industry (acquiring a certain technology, including a certain technological improvement in the products of the company, mastering a new production process, etc.). It has also to meet more generic goals such as understanding a certain chemical process, mastering a certain physical phenomenon and so on. Partners of a joint laboratory know that it requires important means in terms of financing and personnel and long-term horizons beyond the scope of a three-year contract. It is therefore a question of creating an organization that can raise funds, especially local public funds (cities, counties, regions) and European funding in areas that benefit from planning and development policies. In the same way, these organizations can receive young European post-doctoral researchers. Thus the means obtained are sufficient to form a team of 10 researchers on the average (including PhD students) on the same site and the purchase of expensive material.

The CNRS/companies joint research units have their own budget, contrary to industrial contracts that the teams manage within the global budget of the laboratory. This makes it much easier to recruit and pay employees and to buy equipment.

In most cases the joint laboratory is set up on the site of its activities: it can be reserved inside the buildings of the laboratory, the industrialist or on a new site created for the unit. There are strong arguments concerning the advantages of the proximity of personnel and the resulting knowledge externalities: 'by working together every day, public and private researchers exchange ideas and knowledge and come up with new ideas'. We can find, to some extent, the 'myth of the cafeteria' in this line of reasoning. Indeed, it has now been demonstrated that proximity of economic players and more specifically laboratories and companies is due more to practical reasons concerning the commuting of scientific workers (Saxenian 1994) or derives from the spatial form of their social networks (Grossetti and Bès 2001). It appears that the presence on the same site of different researchers enables a better management of the activities by the laboratory head: in contact with the personnel every day, he/she leads the activities and directs the team. The location of the PhD students can moreover be a source of conflict between their scientific director and their manager in the industry.

Moreover, the necessity to carry out work on the same site seems to be more closely linked to the everyday use of test facilities and specific apparatus by the research engineers, the researchers and especially, the doctoral students. This is the instrumental dimension associated with technical competencies that justifies the creation of a specific place dedicated to joint research. The staff waste less time moving about to carry out experiments and the research materials (products, substances, samples, and so on) circulate more rapidly and easily between the teams; this facilitates more integration and protection of knowledge. This is in line with Adam Smith's assessment of the gains from specialization in production.

For that matter, in the case where the joint laboratory is not set up on a single site, but divided into two areas, one for the public lab and the other for the industrialist, researchers and especially PhD students spend a lot of time going back and forth. In this configuration we can note that the scientific equipment set up in the laboratory and on the industrial site are relatively different: public researchers will buy measurement tools and tests common to several other projects while engineers and industrial researchers will purchase equipment closely related to the production process or their product.

However, new human resources in science and technology (Bozeman 2000) are created in these joint laboratories, if only through the training of doctoral students and the recruitment of some of them. Cassier (2002) underlines the fact that the creation of human resources in technologies that directly interest the industrialist is probably one of the most important results of a joint laboratory. He quotes the example of a joint laboratory that progressively led to the creation of an internal research department in a company and the progressive hiring of doctoral students trained in this organization. Therefore, this procedure accustoms young researchers to work on collaborative projects.

Concerning the monitoring of activities, there is better knowledge upstream of the work of each partner and therefore a stronger mutual understanding. For example, we can observe a better control over the use by researchers of the technologies, know-how and products of the industrialist and the eventual publication of information on the research carried out, above all concerning the characteristics of the products or industrial processes. The industrialist exerts intense pressure on the researchers and engineers to prevent knowledge, in the broad sense of the word, from escaping from the laboratories. But with this type of organization, they control even better the communication of research activities because the public laboratory is located within the company's walls: the members of the joint laboratory control visits, the inflow/outflow of materials, the movements of the researchers and the optimization of discoveries. Moreover, these joint research projects mobilize public researchers on a full-time basis and they have less time to compare their results with some of their colleagues on other applications or to launch freer, more risky, projects. They temporarily lose control of their agendas but in return have more means to carry out their work.

For public researchers, working closely with industrialists enables them to be more productive because they work directly with the technologies and the know-how of the companies that are otherwise highly confidential and protected. However, when shared research concerns products, public researchers are hampered in their scientific activities – the company refuses to reveal all the characteristics of the product – and also in publishing their work because they must not inform potential competitors. An important aspect in the control of activities relates to the ownership and the exploitation of results. In joint laboratories associated with the CNRS, the rules of ownership are strict but not specific: the industrialist has exclusive rights concerning a field of application or a technological field that can be relatively vast in relation to the scientific area of the laboratory. Here, we can note that joint laboratories often include private database projects such as combinational libraries, natural products libraries or private libraries. This aspect linked to the control of scientific databases is vital in the relations between public research and industry and touches on a wider aspect of the political economy of patents, that of the appropriation of public property, e.g. literary and artistic works or live performances (Lévêque and Menière 2003).

3. COMPLEMENTARITY AND SUBSTITUTABILITY OF ORGANIZATIONAL FORMS FOR COLLABORATIVE RESEARCH

Joint laboratories are not created *ex nihilo* by scientific partners to make an occasional 'good deal': they are consequential to prior commitments between members, at the level of either the teams or the laboratories. People in charge know each other before creating a mixed organization. But the creation of this type of integrated organization is neither inevitable nor irreversible.

In this last part, we will examine the question as to whether the joint laboratory is necessarily the final form of a prior relationship between an industrialist and a laboratory. Is there a framework that is progressively set up and that leads quasi-systematically from the hiring of researchers (through research contracts and specialized assignments) to a collective commitment reinforced by the creation of a joint laboratory? This could arise if there were a sort of interpersonal economy created first by leaders, then growing up through the different levels of the activity concerned from the individual to the team, and its laboratory, up to the supervisory institutions.

In the case of these integrated forms of research, the answer is negative:

- On the one hand the individual interventions of public researchers (expertise, consulting, exchanges, etc.) often take place after a first series of contracts and more often at the end of a project when the company invests less money in contracts. We can observe the existence of a prior confidence in the research team and then the creation of privileged relations with the researcher in particular. The joint laboratory is not the result of a series of prior commitments that are stronger and stronger. Its construction is based on another logic, driven by issues of intellectual property and the sharing of means.
- On the other hand, companies do not limit their collaborative projects to joint laboratories: they simultaneously maintain relations with other national and foreign research teams. They therefore have an interest in varying contractual commitments during a collaborative project but also during a period of intensive research and development activity.

By mapping the dynamics of the levels of activity invested in science/industry collaborations we obtain two ideal/typical modes:

Mode 1: stacking of commitments

Contract with one team → team + researchers → team + researchers + second team → team + researchers + joint laboratory, and so on . . .

Mode 2: alternating levels of commitments

Contract with one researcher → contract with one team → contract with one researcher → joint laboratory → contract with one team → contract with one researcher → framework agreement, and so on . . .

Considering the recent creation of CNRS joint laboratories, it is necessary to refine these hypotheses through further studies on the outcome of the commitments of the partners.

4. CONCLUSION

Because we are dealing with relations between science and industry we can note an intensification of debates around the theme of the commercialization of knowledge (Weingart 1997), which are particularly heated in France as regards the patenting of live performances, copyright or free software.

But the current conflicts that exist on intellectual property rights for laboratory specifications or the localization of doctoral students co-sponsored by industry are the signs of a transfer of the control of research results towards the phases of the activity itself. In fact the CNRS has strengthened its demands in terms of intellectual property rights. The stakes currently concern scientific and technical expertise as illustrated by the increase in the number of technological platforms created in the laboratories, which are aware of their specific advantages in terms of scientific and technological equipment and often propose the sale of technological services that are disconnected from the management of research projects.

Since we are dealing with joint public/private laboratories, there is of course an important stake involving the control of the scientific knowledge obtained and its externalities, but also the insertion of these projects into wider scientific networks (networks in companies, laboratories) to ensure the circulation of results beyond the barriers erected by contracts of association. The expression 'walls have ears' seems particularly appropriate for these organizations.

NOTES

1. Our empirical data were gathered during four surveys: the first was carried out with Michel Grossetti in 1999–2000 on the origins of the 133 collaborations that took place during the period 1960–2000, between the engineering science teams of the CNRS and industrialists and involved long and semi-directive interviews of researchers, engineers and technicians (140). A second phase, that I carried out alone, concerned the dynamics of the collaborations between laboratories and industrialists in two specific scientific areas (porous media and electrical engineering) and took place in 2001–02. In addition to the interviews this phase was also based on the CVs of the researchers (30) and laboratory reports (32). I launched a third phase in 2003–04 that concerned more specifically the work of doctoral students and laboratory technicians: the empirical data consisted in detailed accounts of the 21 industrial theses by the doctoral students and their supervisors but also various material traces of the theses (articles, notes on experiments, CVs, papers for symposia, articles in the press, photos of experiments, and so on) and reports on visits to laboratories and companies (15). In 2005–06, we visited the joint CNRS/industry laboratories (7) and interviewed their directors as well as the researchers of these organizations.
2. We have demonstrated that this asymmetry of information can be the cause of the creation of specific scientific assets (Bès 2005b).
3. For more details see Bès (2005a).

BIBLIOGRAPHY

Bès, M.-P. (2004), 'Connaissances et relations sociales des jeunes chercheurs', *Recherches sociologiques*, **XXXV** (3), 123–35.

Bès, M.-P. (2005a), 'La diversité des partenariats Universités-entreprises: formes organisationnelles, nature des recherches et mobilité des personnels', paper presented to the *RESUP* [Réseau d'Etude sur l'Enseignement Supérieur], Recherche universitaire et enterprises seminar, Sciences Politiques, Paris, 9 June, accessed 10 May, 2007 at www.resup.u-bordeaux2.fr/.

Bès, M.-P. (2005b), 'Savoir et savoir-faire élaborés dans les relations science-industrie: un nouvel enjeu pour la marchandisation de la science?', *Sciences de la Société*, **66**, (October), 131–47.

Bozeman, B. (2000), 'Technology transfer and public policy: a review of research and theory', *Research Policy*, **29** (4–5), 627–55.

Bozeman, B. and V. Mangematin (2004), 'Editor's introduction: building and deploying scientific and technical capital', *Research Policy*, **33** (4), 565–8.

Callon, M. (1997), 'Analysis of strategic relations between firms and university laboratories', presentation to the conference on the Need for a New Economics of Science, University of Notre-Dame, 13–16 March.

Cassier, M. (2002), 'Private property, collective property, and public property in the age of genomics', *International Social Science Journal*, **171**, 83–98.

Cassier, M. and D. Foray (2002), 'Public knowledge, private property and economics of high-tech consortia: case studies in biomedical research', in *The Dynamics of Inter-firm Cooperation: A Diversity of Theories and Empirical Approaches*, Cheltenham, UK and Northampton, MA, USA: Edward Elgar, pp. 192–3.

David, P. (2003), 'The economic logic of "Open Science" and the balance between private property rights and the public domain in scientific data and information: a primer', in National Research Council, *The Role of the Public Domain in Scientific Data and Information*, Washington, DC: National Academy Press.

Duhautois, R. and S. Maublanc (2005), 'Les carrières des chercheurs dans les entreprises privées', Centre d'Etudes et de l'Emploi research report no 25, September, accessed 10 May, 2007 at www.cee-recherche.fr/.

Dumont, B. and P. Holmes (2002), 'The breadth of intellectual property rights and their interface with competition law and policy: divergent paths to the same goal', *Economics of Innovation and New Technology*, **11** (2), 149–62.

Foray, D. (2004), *The Economics of Knowledge*, Cambridge, MA: MIT Press.

Grossetti, M. and M.-P. Bès (2001), 'Interacting individuals and organizations: a case study on cooperations between firms and research laboratories', in A. Kirman, and J.-B. Zimmermann (eds), *Economics with Heterogeneous Interacting Agents*, Guildford: Springer, pp. 287–302.

Joly, P.B. and V. Mangematin (1996), 'Profile of laboratories, industrial partnerships and organization of R&D: the dynamics of relations with industry in a large research organization', *Research Policy*, **25** (4), 901–22.

Latour, B. (1987), *Science in Action: How to Follow Scientists and Engineers Through Society*, Cambridge, MA: Harvard University Press.

Levêque, F. and Y. Menière (2003), *Economie de la Propriété industrielle*, Repères, Paris: La Découverte.

Mangematin, V. and S. Robin (2003), 'The double face of PhD students: the example of life sciences in France', *Science and Public Policy*, **30** (6), 405–14.

Mustar, P. and P. Laredo (2002), 'Innovation and research policy in France (1980–2000) or the disappearance of the Colbertist state', *Research Policy*, **31** (1), 55–72.

Nelson, R.R. and S.G. Winter (1982), *An Evolutionary Theory of Economic Change*, Cambridge, MA: Harvard University Press.

Polanyi, M. (1958), *Personal Knowledge: Towards a Post-critical Philosophy*, Chicago, IL: University of Chicago.

Powell, W.W. (1990), 'Neither markets nor hierarchy: networks of organization', *Research in Organizational Behavior*, **12**, 295–336.

Romer, P. (1986), 'Endogenous technical change', *Journal of Political Economy*, **94**, 1002–37.

Saxenian, A. (1994), Regional Advantage, Culture and Competition in Silicon Valley and Route 128, Cambridge, MA: Harvard University Press.

Weingart, P. (1997), 'From finalization to mode 2: old wine in new bottles?', *Social Science Information*, **34** (4), 591–613.

3. Proprietary vs. open-access dimensions of knowledge

Marc Isabelle

1. INTRODUCTION

Public research systems have experienced deep transformations worldwide since the 1980s in connection with the competitiveness agenda. In European countries, they are now strongly challenged by the Lisbon strategy of the knowledge economy (a modern version of the competitiveness agenda) as prominent engines of knowledge production and dissemination. Among the changes that have taken place, researchers are advised to incorporate more consideration of economic and societal needs in their activities, as well as providing more protection for their results and working in closer connection with private firms (Ziman 1994; David 2000). Some scholars advocate a new social contract for science on this basis (Gibbons 1999), whereas others have expressed concerns about the pitfalls that may arise from this shift (Florida and Cohen 1999; Florida 1999; David 2003).

In the context of increased attention paid to technological innovation as a driver of economic growth, it seems quite obvious why one would like the public research sector to work more interactively with firms and the society at large: public research organizations (PROs) would be better aligned with and more responsive to firms' needs, the knowledge and technologies they develop would be transferred more rapidly and effectively to firms, etc. However, for more subtle reasons, the greater relevance of public research may inadvertently produce effects that have negative impacts on the bottom-line of the social rate of innovation. These negative consequences can be understood in terms of departures from a normative model of how the scientific system and the market economy ought to work together (Nelson 2004).

In this model, the function of the scientific system is essentially to build a knowledge base that is largely and freely accessible to business firms and from which they can draw for their innovative activities. This function is satisfied through a set of rules, norms and reward systems that altogether characterise the 'Republic of Science' (Polanyi 1962; Merton 1973). The rapid publication of research results is particularly important, to enable their use and evaluation by others and ground the cumulativeness of the scientific enterprise. Another is that the definition of the research agenda should take into consideration only scientific goals and not practical outcomes. As a matter of fact, the economic payoffs of basic research are largely unpredictable and it would be misleading to initiate or conversely to reject a project according to its economic (ir)relevance. Incorporating more considerations for economic and societal needs encroaches upon this point, while the appropriation of results (e.g. through patents) jeopardises the scientific commons.

Concerning the alleged tension between basic research and economic relevance, Donald Stokes's contribution of 'Pasteur's Quadrant' (Stokes 1997) pushed the debate forward by showing that basic research need not be curiosity-driven but could be use-inspired – and actually was in many cases. More generally, the disconnection of fundamental research from economic and societal concerns, captured by the metaphor of the ivory tower, could well have been overestimated as well as idealised by the proponents of the linear model of science-based innovation (Nelson 2000; Brooks 1994; Mowery et al. 2001). One illuminating historical example of use-inspired basic research was that carried out at the Bell Labs in the 1940s in the field of solid-state physics, which had the applied purpose of searching for better amplifiers of telephone signals and eventually led to the invention of the transistor (Brooks 1994).

However, Stokes's 2D taxonomy does not say anything about the tension between the scientific commons and the growing importance of appropriation mechanisms in the sphere of public research. As it misses this essential dimension of the normative model sketched above, this taxonomy cannot function properly as a tool for characterising research activities in the emerging framework of the knowledge economy. For this purpose, it has to be completed with a third dimension, that of proprietary vs. open-access research. Note that this additional dimension does not refer to the intrinsic properties of the activities contemplated but is more broadly related to their social organisations in terms of the objectives that are deemed legitimate by those engaged in research, of their reward system, of the modes of disclosure and the regime of appropriability of the results (Dasgupta and David 1994).

In this chapter, a new taxonomy that expands Pasteur's Quadrant with the proprietary vs. open-access dimension will first be presented and discussed. An important question immediately follows the design of the expanded taxonomy: along which of its dimensions does the global pressure for increased relevance of research unfold? The chapter will report the results of a survey that used the new taxonomy to analyse the changes that are occurring in a large public research organisation: the French Commissariat à l'Energie Atomique (CEA). The CEA is of special interest as compared to its peers in France and Europe because its activities are widely distributed across the various dimensions of the new taxonomy. Its missions are essentially technological but many research activities are triggered by curiosity-driven factors and pursue fundamental understanding objectives on top of developing technical devices. Moreover, it has a consistent track record of patent applications and first-class scientific publications. Consequently, the CEA appears to be a propitious organisation in which to probe the expanded taxonomy.

The motivation to review and expand the existing definitions and typologies of research activities was first stimulated by the researchers and the management at the CEA themselves. In fact, during a series of presentations of a study in progress dedicated to analysing the competitive advantage that the CEA could leverage from its ability for internally coupling scientific and technological research, the audience repeatedly mentioned their felt need for precise definitions of the various types of research performed at the CEA. They almost unanimously agreed that the 'fundamental' vs. 'applied' research distinction was no longer suitable and preferred the terminology of scientific vs. technological research, but asked for more in-depth information about what 'scientific' and 'technological' really meant.

The remainder of the chapter is organised as follows. Section 2 reviews the various characterisations of research activities that are found in the economic literature. It also points to the need for an expanded taxonomy which is then defined and discussed. Section 3 reviews the literature about how the increased pressure for economic relevance manifests itself in the world of public research organisations. Section 4 analyses this issue on the empirical basis of a survey conducted within the CEA with the expanded taxonomy. Section 5 presents our conclusions.

2. CHARACTERISING RESEARCH ACTIVITIES

The identification of a limited set of economically relevant dimensions of research activities, for classification or modelling purposes, is a task that has long focused the efforts of many scholars. After first reviewing the existing literature in this field, a new taxonomy will be presented and discussed, which builds upon pre-existing work but also adds to it by taking into account the profound changes that research activities have recently experienced in terms of knowledge appropriation.

2.1 Existing Definitions and Taxonomies

A pivotal and historically-rooted question concerning the economics of research activities is their degree of exogeneity, i.e. to what extent is the research sector interconnected with the economy and through which channels? At one extreme, in the linear model view, the research sector is seen as functioning autonomously and simply feeding the economy with new scientific and technological knowledge or information. On the other hand, many departures from this exogenous model are possible, for example, the financing of the research sector comes from the economy and may fluctuate with it, some issues tackled by the research sector may have originated in the stream of economic activities, or some research activities may be deliberately carried out to produce competitive advantages for specific economic actors.

These various forms of the exo/endogeneity of research activities have generated a mushrooming terminology. A distinction that is repeatedly mentioned is that of 'basic research' as opposed to 'applied research' but these terms also include 'fundamental research', 'curiosity-driven research', 'blue-sky research', 'use-inspired research', 'upstream research', 'scientific research' and 'technological research'. These terms are seldom defined with enough precision. However, there have been several attempts by scholars or official bodies to provide careful definitions of some of these terms, to question their signification or to build taxonomies of research activities.

Calvert and Martin (2001) use a survey of UK and US scientists to analyse in detail the different meanings of 'basic research', starting from the point that the term is often invoked but the concept is rarely analysed, leading to confusion. They find most interviewees have an epistemological definition in mind when using the term 'basic research', i.e. they refer to the unpredictability of research, to its generality, to the theory and reductionism behind it, as well as an intentional one, i.e. basic research is driven purely by curiosity. Only half as common is the reference to distance from applications and only a quarter is the reference to the institution where the research is done (i.e. basic research is university research) as well as the norms of disclosure (i.e. basic research is published research). The least frequent association is with specific scientific fields that are implicitly considered to grow out of basic research.

Official definitions of the *Frascati Manual* (OECD 1994) oppose 'basic research' to 'applied research'. They rely on the epistemological, the intentional as well as the norms of disclosure references but omit the distance from application, the institutional and the scientific field references that some researchers have in mind. Thus, according to the OECD, applied research is defined as being directed primarily towards a specific practical aim or objective and its results are intended primarily to be valid for a single or limited number of products, operations, methods, or systems. Lastly, the knowledge or information derived from it is often patented but may also be kept secret. Conversely, the OECD considers basic research to be experimental or theoretical work undertaken primarily to acquire new knowledge of the underlying foundations of phenomena and observable facts, without any particular application or use in view. It analyses properties, structures and relationships with a view to formulating and testing hypotheses, theories or laws. Furthermore, the results of basic research are not generally sold but are usually published in scientific journals or circulated to interested colleagues.

From these definitions, it clearly appears that basic and applied research are multidimensional concepts. Hence, asking researchers whether the changes in their research environment has induced a shift towards more 'applied research' as opposed to 'basic research' will not tell us much about the true nature of the shift: does it mean less theory or generality, or less room for curiosity-driven activities, or activities that are closer to applications, or more and more patented results? Introducing additional subdivisions along the linear basic/applied research axis (e.g. 'applied technology', 'technological research and engineering', 'targeted scientific research' and 'basic scientific research' as defined by Narin et al. 1976) does not help much in this respect.[1] This chapter is an attempt to disentangle unidimensional conceptions of research activities that will be combined in a new taxonomy with the purpose of better understanding in which directions the pressure for increased relevance really goes.

This methodological point of view has already been adopted by Stokes (1997). Departing from the linear model of science-based innovations that emanated from Vannevar Bush's influential report to the US President Roosevelt (Bush 1945), Stokes opposed a widespread opinion according to which research activities intended to be useful were presumably pointless in terms of fundamental knowledge generation. Building on his investigation of Louis Pasteur's research experience, he recommended that the fundamentality of research activities on the one hand (the epistemological dimension) and their utility on the other hand (the intentional dimension) were two separate dimensions and he concluded that basic research (i.e. research targeted towards fundamental understanding) could be use-inspired and actually was in many cases. Stokes's 2D taxonomy (see Figure 3.1) was widely adopted as a tool for characterising research activities and occasionally for shaping discourses about science and technology policy. In connection with this, Pasteur's quadrant has been extensively used to refer to 'use-inspired basic research'.

Considerations of use?

Quest for fundamental understanding?	No	Yes
Yes	Pure basic research (Bohr)	Use-inspired basic research (Pasteur)
No		Pure applied research (Edison)

Figure 3.1 Stokes's 2D taxonomy

2.2 An Expanded Taxonomy: Motivation and Presentation

Useful as it may be, Stokes's 2D taxonomy does not say anything about the ownership and control a research institution or a researcher might have over the results of their activities. Yet this dimension of public research activities has brought increasing attention from scholars and policy-makers following the reforms of public research systems of the past 25 years. These reforms first took place in the USA with the Bayh-Dole act of 1980 and then diffused to Europe. In France, for example, a major law on innovation and research was passed in 1999, with the double purpose of stimulating technological innovation and economic growth at the national level as well as increasing economic returns from investment in public R&D (see, for example, Cohen et al. 1998; OECD 1996). To this end, most of them strongly promote the appropriation of their research results by PROs – essentially through patenting – and their licensing to business firms.

However, the development of intellectual property instruments in the public research sector is highly controversial. In point of fact, some scholars have stressed the importance of public research systems delivering freely accessible knowledge and building a knowledge-base for national innovative capacity (Nelson 2004). They have also pointed to the possible damage that a broad diffusion of the patenting of results emanating from public research may induce on innovation, through the fragmentation of the knowledge base (Heller and Eisenberg 1998). So the proprietary vs. open-access dimension is essential when it comes to fully grasping the economics of how the research sector could foster the knowledge economy. It may be that use-inspired basic research *à la Pasteur* has its results either published or patented, and this makes a big difference in the way these results could be used for other research or for innovation purposes, and eventually from the social welfare point of view.

Moreover, Dasgupta and David (1994) have argued that the behaviour of researchers in terms of the appropriation or publication of the results that proceed from their research lies at the core of the distinction between the scientific system and the technological system, which itself grounds the normative model presented in the introduction. More generally speaking, they contend that the social organisations that prevail in each research system are more relevant to distinguishing between communities of researchers than their methods of enquiry, the nature of the knowledge obtained or the sources of their financial support. Notice however that their definition of 'the Realm of Technology' and 'the Republic of Science' are normative in the sense that they bind together the various dimensions, e.g. research in the republic of science is meant to be curiosity-driven, to produce new fundamental knowledge and to be published. Conversely, a researcher working on a project for a firm and publishing some results might not find his or her place in either of the two systems.

Following the previous discussion, it appears that three disentangled dimensions of the research space should be retained. The first relates to the epistemological category of Calvert and Martin (2001) and is concerned with the very nature of the research activities. On the one hand, there are research activities that produce 'fundamental knowledge', i.e. new understanding of the laws of nature based on theory building, reductionism and modelling. On the other hand, there are research activities that yield 'technical achievements', for which understanding is an issue insofar as it helps in getting things to work. The second dimension is the intentional one: 'curiosity-driven' research is undertaken to satisfy the curiosity and interest of the individual researcher and of the research community at large, while 'use-inspired' research is undertaken with the expectation that it will be useful. Lastly, the 'open-access' vs. 'proprietary' distinction refers to the norms of appropriation of the results, of control over them and also of their disclosure (see Figure 3.2). Basically, research results are published in journals and made open-access in the first regime whereas in the second they are appropriated essentially through patenting (or by being kept secret).

2.3 An Expanded Taxonomy: Discussion

Note that the axes in this expanded taxonomy have a different interpretation than in Stokes's 2D taxonomy. In the latter, the opposite ends of the axes express a yes/no antagonism with the consequence that the taxonomy effectively exhibits four quadrants. In the expanded taxonomy, the opposite ends of any axis are not necessarily antagonistic. As my survey of researchers within the CEA will reveal, some research activities may generate new fundamental knowledge and at the same time lead to technical

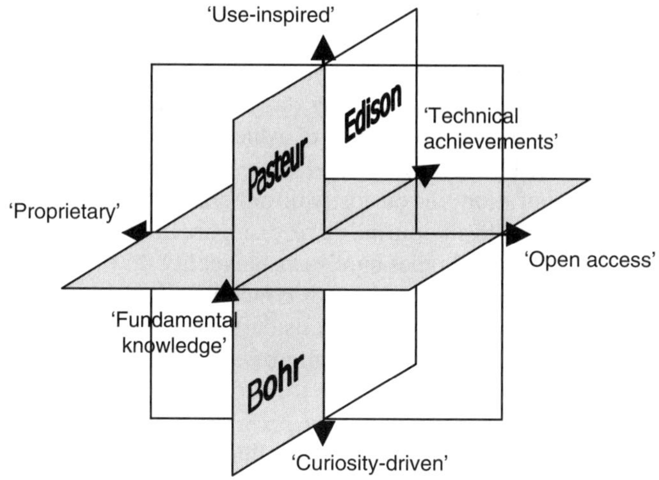

Figure 3.2 Disentangling three dimensions of the research space

achievements. Some may also be driven by curiosity in the sense that they are considered as areas worth exploring by scientific communities while they may also be expected to yield useful results. Contrary to what Stokes's taxonomy suggests, there does not seem to be any reason why the sets of research activities triggered by researchers' curiosity on the one hand and by the expectation to produce useful outcomes on the other hand could not overlap. Finally, taking out a patent on an invention does not prevent researchers from publishing their results in academic journals afterwards.

At the end of the day, the expanded taxonomy gives rise to $3^3=27$ different options for characterising research activities, so it provides a much more detailed characterisation than Stokes's 2D taxonomy. Given the non-opposition between both ends of any axis, the 3D 'xyz' representation is probably not best suited to serve as a tool for representing a given research activity, a purpose for which a 'radar' representation would fit better. However, I choose to represent the taxonomy as it is to illustrate that it expands Stokes's 2D taxonomy. Moreover, this representation is consistent with the fact that each axis represents a homogeneous dimension of the research space, i.e. (1) the motivation for the research, (2) the nature of its outcomes, and (3) the way its results are appropriated, controlled and disclosed.

Of course, the three dimensions may be correlated in practice although they are disentangled from the logical point of view. For example, idealistic types of research activities easily find their way into the new taxonomy:

'fundamental research' refers to curiosity-driven research activities that yield new fundamental knowledge and have their results published in academic journals without paying attention to appropriation, technical achievements or expectations of usefulness. Conversely, 'applied research' stands for use-inspired research activities, the results of which are technical achievements that are appropriated by their inventor. For this type of research, fundamental knowledge elaboration, the curiosity of any stakeholder and the publication of research results in journals are disregarded. It is however most probable that these stylised types barely exist in reality.

As far as the intentional dimension is concerned, it is worth mentioning an *ex post–ex ante* difference. That considerations of use do (or do not) contribute to triggering the undertaking of a particular research activity does not imply that the research will (or will not) be useful *ex post*. As already mentioned, it is in the nature of research activities to be subject to high uncertainty. Some research activities that intend to be useful end up not that useful, and conversely others end up with a wide range of applications although they were by and large driven by curiosity (e.g. the discovery of radioactivity). What matters for the purpose of this chapter is the *intention* of the researcher or research team in charge of the activities contemplated with respect to encompassing considerations of use, be they *ex ante* or *ex post*.[2] Finally, one should note that some research activities may move across the taxonomy as time passes, for example, go from an essentially (curiosity-driven – fundamental knowledge – open-access) position to one that is more (use-inspired – fundamental knowledge – proprietary).

3. WHAT DO WE KNOW ABOUT INCREASED PRESSURE FOR THE RELEVANCE OF RESEARCH?

An abundant literature in the past decade has studied the effects of the increased pressure for the relevance of public research (as referred to by Pavitt 2001), but the results are still controversial. The main theoretical and empirical results related to the skewing problem, the tragedy of the anticommons and the secrecy problem will now be presented in turn.

3.1 The Skewing Problem

A first concern is about the threat to basic research that may arise from a shift to more short-term and applied research in response to the requirements of industrial partners, known as the skewing problem (Florida and Cohen 1999). The 'short-term' and 'applied' terms refer to the lack of room for curiosity-driven research as well as an attitude that might be incompatible with fundamental knowledge generation. Hence, the skewing problem has connections with two dimensions of the expanded taxonomy.

From the economic point of view, the skewing problem relates to the inefficiencies that may stem from the allocation of resources for research according to their anticipated economic returns, a feature that is clearly related to use-inspiration in the expanded taxonomy. The problem is that the market economy is considered a poor mechanism to allocate resources for scientific research given the serendipity of its results. Conversely, these resources would be best allocated when informed scientists themselves decide what the important subjects for scientific inquiry are (Nelson 2004). This point clearly calls for a balance between curiosity-driven and use-inspired research, on which the various sources of financial support that are available for researchers have a great impact. Typically, recurring funding allows scientists and communities to select research themes according to their own judgement while industrial or government agency funding selects a priori research themes that are anticipated to yield practical payoffs.

The empirical evidence on these points does not deliver a definite answer. Blumenthal et al. (1986) found that researchers whose activities were supported by industry were four times more likely than others to report that their decisions to explore a particular research topic had been impacted by the expectation that it would yield commercial applications. Rahm and Morgan (two authors cited in Florida and Cohen 1999) find a small positive relation between the involvement of academia with the industry and increased levels of applied research, although the direction of this relation remains open to question.

Goddard and Isabelle (2006a) found on the basis of a survey of 130 laboratory directors in French PROs that, because of collaboration, labs tend to step up their 'applied research and experimental development' and, to a lesser extent, their 'oriented basic research' while this shift is only marginally accompanied by a diminution in 'pure basic research'. Addressing more specifically the issue of the returns to public R&D, Henderson et al. (1998) use American universities patent citation counts to show that some measures of the generality and of the importance of these patents have decreased since the early 1980s and that correlatively, the strong increase in university patenting triggered by the Bayh–Dole act of 1980 encompasses essentially low-quality patents.

By contrast, Ranga et al. (2003) find on the basis of publication classification and counts at the Belgian Katholiek Universiteit Leuven that there is no evidence of a shift towards applied research in connection with the involvement in university–industry linkages. In an extension of this work, Van Looy et al. (2004) compare scientists at the Katholieke

Universiteit Leuven who engage in contract research with firms to those who do not, and find that the first group publishes more in applied fields without publishing less in the basic fields. Hicks and Hamilton (1999) also studied a very large sample of university publications and university-industry co-publications over the 1981-94 period and found that the proportion of basic research papers did not decrease over time, although co-authored papers between academia and researchers in firms did increase.

Two other concerns with the increased relevance of public research relate to the third dimension of the expanded taxonomy, i.e. the appropriation, control and disclosure of results.

3.2 The Tragedy of the Anti-commons

First, the patenting of research results may have detrimental effects on innovation. Heller and Eisenberg (1998) have shown that the fragmentation of the scientific and technological knowledge base could lead to a tragedy of the anti-commons, in the sense that the high coordination costs that must be incurred in order to gather complementary inventions scattered among many different owners, as well as the strategic behaviour of the latter, may delay or simply make it unprofitable to develop an innovation that would otherwise have been developed, that is, had the inventions been freely available in the public domain. In the aggregate, it is the whole innovation activity itself that could be deterred. This risk is more prevalent with complex and integrative technologies (Shapiro 2001) or where the innovation process is highly cumulative (Scotchmer 1991).

It is only recently that several studies have been implemented to yield empirical evidence on this point, with different focuses and methodologies. Walsh et al. (2005) surveyed US academic scientists in the field of biomedical research in order to ascertain the extent of the 'patent thicket' in upstream research. They find that patents on research inputs do not divert researchers from pursuing specific projects and only very marginally delay or modify them, one apparent reason for this being that academic scientists are barely aware of the patents that are relevant to their research. It may well be that industry scientists would tell a different story.

Murray and Stern (2006) take a different perspective and find more pessimistic results. They directly investigate the anti-commons in scientific knowledge diffusion by focusing on paper-patent pairs[3] and find a moderate drop in a paper's citations after the associated patent is granted (more than three years later on average). They conclude that a modest anti-commons effect does exist in the sense that published findings that are associated with formal intellectual property rights have a lower cumulative scientific impact than in the absence of such IPRs.

3.3 The Secrecy Problem

The secrecy problem (Florida and Cohen 1999) relates to restrictions over the disclosure of research results that may be imposed by industrial partners. The question of 'what can (not) be published and when?' clearly relates to the 'open-access' dimension of the expanded taxonomy. It also connects to the 'proprietary' dimension in the sense that publications may be delayed to allow for patent application. These restrictions have been extensively documented in the literature.

Blumenthal et al. (1997) surveyed a large sample of scientists in the field of life science and found that participation in collaborations with private firms and commitment to the commercialisation of research results were often associated with publication delays, and that the main drive behind these delays was to allow time for filing patent applications. Cohen et al. (1994) surveyed 1056 joint university–industry research centres in the USA and found that for half of them, the industrial partner could force a delay in the publication of research results while for one third, it could have some information deleted from papers prior to publication. Exploiting a similar survey for French PROs, Goddard and Isabelle (2006b) find comparable figures, with the additional result that 26 per cent of respondents effectively experienced restricted information from publications.

Delayed or precluded publication is not the only form that restrictions on scientific communication can take. In fact, informal exchange between researchers is another important vector for the diffusion of scientific information, and the decision to share information informally with colleagues or on the contrary to withhold that information is another difference captured by the 'open-access' vs. 'proprietary' dimension that the present taxonomy adds to Pasteur's Quadrant. Here again, the literature does not offer definitive answers.

The survey by Blumenthal et al. (1997) did not find any significant relation between the involvement of researchers with industry and the withholding of research results. In their comparison of two distant surveys of mathematicians, physicists and experimental biologists conducted in 1966 and 1998, Walsh and Hong (2003) confirm an increased practice of secrecy among academic scientists and analyse the predictors of this trend. They find a mixed effect of the scientists' commercial activity, given that secrecy increases with industry funding but decreases with industry collaborators.

The survey of lab directors in French PROs analysed by Goddard and Isabelle (2006b) directly addresses this question of the limitations on scientific communication induced by collaboration with industry. Not surprisingly, a majority of respondents report limitations with firms other than the partner(s) but 30 per cent also experience restrictions with researchers in other PROs. A striking result is that limitations even affect communication within the lab (although to a lesser extent), whereby it is the true locus of scientific and technological knowledge production that is torn apart.

4. EMPIRICAL RESULTS

In this section, the expanded taxonomy is used to analyse empirically the changes that are occurring in the type of research performed at a large public research organisation: the French Commissariat à l'Energie Atomique. I will present the CEA and explain why it is a suitable organisation to test the new taxonomy of research activities, before turning to the description of the preliminary survey and its results.

4.1 Why Probe the Expanded Taxonomy in the CEA?

A characteristic of French public research stems from its duality in the sense that universities coexist in approximate parity with government laboratories ('*grands organismes*' as they are called in France) that, as opposed to the former, are mostly oriented towards technological research in specific areas (medicine, energy, defence, ICT, etc.). The Commissariat à l'Energie Atomique is the largest of these government labs and the second largest public research organisation in France by number of employees.[4] It has distinctive features that make it propitious for probing the expanded taxonomy. Its activities are distributed over a large bandwidth on the research spectrum – from fundamental to applied research, to keep referring to the traditional terminology – in some targeted technological areas: defence and global security, energy, information and health technologies.

First, what gave birth to the CEA and subsequently defined it best are its technological missions, often related to national strategic objectives: providing France with nuclear weaponry and developing the technological infrastructure for the French electronuclear industry (from the time of its inception in 1945); fostering technological innovations in the fields of ICT and new technologies for health, since the 1980s. From this point of view, research at the CEA tends to be mostly use-inspired and to a large extent aiming at technical achievements. At the same time, the CEA has a long experience of working with industry and the patent culture is more diffused there than in most other French public research organisations.[5]

But in addition, the CEA's technological missions rely on selected fundamental research activities that have always been preserved. This is reflected in the CEA's organisation into four operational divisions: Nuclear Energy, Military Applications, Technological Research and Fundamental Research, the latter being composed of two subdivisions: Life Sciences and Matter Sciences. Moreover, the CEA has an important track of scientific records in the fields of, for instance, nuclear physics, astrophysics, particle physics, thermonuclear fusion, and also in many domains of chemistry and biology. On top of these complementary missions, the CEA has an established technological capacity for the design, construction and operation of large scientific instruments, from thermonuclear fusion reactors[6] to particle accelerators and satellites.

Another obvious reason to study the CEA specifically is that it can be observed from the inside given that I have been working as an economist in its Department of Economic Studies since March 2003. This position is a privileged one in the sense that it gives the possibility to observe and verify the information that is usually unobservable or unverifiable for the outsider. Moreover, one of my jobs there as an economist is precisely to investigate the origins, mechanisms and consequences of the transformations at stake in public research organisations and to produce targeted, accurate and serviceable information in order to help the researchers and the management anticipate and proactively react to them.

4.2 The Survey and its Results

In January 2006, I implemented a survey prototype directed towards selected researchers within CEA to assess the reliability of the expanded taxonomy and to test whether it could be useful as a tool to identify in which directions the pressure for increased relevance of public research really goes. The sample size was of about 50 individuals and the empirical results presented here are based on the 37 answers received – 37 is not a large number so these prototype results must be interpreted cautiously. However, after controlling for various clustering effects (see below), the variability of data appears to be quite low, so I contend that additional data in the future would probably tend to reinforce the results sketched hereafter. The questionnaire used for the survey is reproduced in Appendix 3.1. Its questions basically stick to the expanded taxonomy in order to characterise the research activities contemplated, the changes they undergo as well as to identify the sources of these changes. Answers were coded on a three-point Likert scale for Q0, Q1 and Q3 and on a five-point Likert scale for Q2 (i.e. as 1, 2, 3, 4, 5 from the leftmost to the rightmost column).

The researchers who were surveyed are outstanding CEA researchers who were appointed 'research director' (henceforth RD). This highly selective grade is specific to the CEA and was instituted in 2000. Awarding it is a priori independent of the researcher's hierarchical position in the organisation but testifies that he/she has achieved major scientific or technological

Table 3.1 Distribution of respondents among the CEA's five operational divisions

DAM	DEN	DRT	DSM	DSV	All together
27%	16%	16%	27%	14%	100%
10	6	6	10	5	37

Note:
DAM = Division des Applications Militaires/Division of Military Applications
DEN = Division de l'Energie Nucleáire/Division of Nuclear Energy
DRT = Division de la Recherche Technologique/Division of Technological Research
DSM = Division des Sciences de la Matière/Division of Matter Sciences
DSV = Division des Sciences de la Vie/Division of Life Sciences

successes during his/her career. Thus, RDs form a network of high-skilled scientific experts whose mission is to take part in the scientific and technical appraisal of CEA programmes, to provide assistance with respect to strategic planning, and lastly to disseminate scientific and technical knowledge at the national and international levels. RDs are promoted for a five-year period and can be re-appointed. In 2005, there were 109 RDs representing a wide range of scientific and technical capabilities.

A subset of about 50 RDs was selected according to two criteria: (1) to be distributed across the different divisions in the CEA so as to represent roughly the number of employees in each division; (2) for any given division, their areas of expertise should be as representative as possible of the variety of specialities. Table 3.1 shows the distribution of the 37 respondents among the CEA's five operational divisions.

As compared to the overall number of employees in each division of the CEA, the sample is biased towards an over-representation of the Division of Matter Sciences (DSM) and an under-representation of the Division of Nuclear Energy. Correlatively, basic research activities are probably over-represented in the sample while applied research activities are under-represented. In the taxonomy, this might induce a bias towards the 'fundamental knowledge', 'curiosity-driven' and 'open-access' types, while the 'technical achievements', 'use-inspired' and 'proprietary' types might be under-represented. So the sample does not mean to be representative of all research activities being conducted in CEA.

A first look at the data is provided by Table 3.2, which exhibits the usual descriptive statistics for a Likert scale-based survey, i.e. the median, the mode, the range and inter-quartile range.

These figures first tell that the respondents are mostly happy with the taxonomy, and this is a most confident result (the range and the inter-quartile range show low variability).

Table 3.2 Usual descriptive statistics associated with Likert scale variables

Q0 Satisfaction						
Median	1.0					
Mode	1.0					
Range	1.0					
Inter-quartile range	1.0					
Q1 Position in taxonomy						
	Technical	Fundamental	Use-inspired	Curiosity	Open-access	Proprietary
Median	1.0	2.0	1.0	2.0	1.0	2.0
Mode	1.0	2.0	1.0	2.0	1.0	2.0
Range	2.0	2.0	1.0	2.0	2.0	2.0
Inter-quartile range	1.0	1.0	0.0	0.3	1.0	1.0
Q2 Changes in position						
	Technical	Fundamental	Use-inspired	Curiosity	Open-access	Proprietary
Median	3.0	3.0	2.0	3.0	3.0	2.0
Mode	3.0	3.0	2.0	3.0	3.0	2.0
Range	3.0	3.0	3.0	3.0	4.0	3.0
Inter-quartile range	1.0	0.0	1.0	1.0	1.0	1.0
Q3 Sources of change						
	Industrial	Sources of	Assessment	Hierarchy	Other	
Median	2.0	1.0	2.0	2.0	3.0	
Mode	3.0	1.0	3.0	3.0	3.0	
Range	2.0	2.0	2.0	2.0	2.0	
Inter-quartile range	1.0	1.0	1.0	2.0	0.0	

With regard to the position of research activities in the taxonomy (Q1), the strongest and most confident result (inter-quartile range = 0) is that respondents are engaged in activities that are essentially use-inspired (median = 1 and mode = 1) despite the bias towards basic research identified earlier. This is not a surprising feature concerning the CEA whose central missions are technological. Yet, another very robust result (inter-quartile range = 0.25) is that these activities are also, to a lower extent, curiosity-driven (median = 2 and mode = 2). What is more, the research activities contemplated seem to be aiming more at technical achievements than at fundamental knowledge generation as well as to be more open-access than proprietary (median = 1 and mode = 1 vs. median = 2 and mode = 2 in each case), although these results are less confident.

The identification of the changes that occurred in the past few years (Q2) shows that the research activities contemplated have become more use-inspired and more proprietary (median = 2 and mode = 2), two results that are confident enough. This must be put in perspective with the fact that curiosity-driven activities seem to be by and large unchanged, as well as activities whose results are made open-access (median = 3 and mode = 3; for the latter, range = 4 indicates more dispersed extreme answers). Finally, no shift has occurred concerning activities aiming at technical achievements or activities aiming at fundamental knowledge generation (median = 3 and mode = 3), the latter result being the most robust of all (inter-quartile range = 0).

Lastly, the answers to the question about the sources of the changes that occurred (if any) reveal that the tracking sources of financial support has the strongest impact (median = 1 and mode = 1), and this result is confident enough. Moreover, when looking at the whole sample, there is no clear evidence that other factors have triggered the changes under scrutiny (median = 2 but mode = 3, i.e. the answer 'No' has the highest response rate).

Medians and modes computed for the whole sample may hide heterogeneous responses of subgroups scattered around them. The distributions of responses are shown in Table 3.3 (the related bar charts are reproduced in Appendix 3.2).

A vast majority of respondents (81 per cent) characterise their research as being essentially use-inspired. At the same time, only four respondents in this subset consider their research to be not curiosity-driven at all.[7] Moreover, two occurrences are found of research activities considered to be both essentially use-inspired and essentially curiosity-driven. So contrary to what Stokes' designation of Pasteur's Quadrant as being 'use-inspired basic research' may suggest, use-inspired and curiosity-driven research are not antagonist. However, they can be considered weak substitutes since the two variables have the same value only in the two instances mentioned above. In other words, a vast majority of researchers (84 per cent) consider their

Table 3.3 Distribution of responses

		1 (%)	2 (%)	3 (%)	4 (%)	5 (%)	NR (%)	Total (%)
Q0 Satisfaction		70	30	0			0	100
Q1 Position in taxonomy	Technical achievements	57	30	8			5	100
	Fundamental knowledge	41	54	5			0	100
	Use-inspired	81	16	0			3	100
	Curiosity-driven	24	62	11			3	100
	Open-access	57	38	5			0	100
	Proprietary	35	49	11			5	100
Q2 Changes in position	Technical achievements	14	32	41	8	0	5	100
	Fundamental knowledge	0	24	59	14	3	0	100
	Use-inspired	14	54	27	3	0	3	100
	Curiosity-driven	0	11	57	27	3	3	100
	Open-access	3	24	57	14	3	0	100
	Proprietary	5	46	30	14	0	5	100
Q3 Sources of change	Industrial partners	22	32	38			8	100
	Sources of financial support	49	32	11			8	100
	Assessment systems	16	35	41			8	100
	Hierarchy	30	27	35			8	100
	Other	16	3	73			8	100

research to be both use-inspired and curiosity-driven but most frequently one type dominates while the other is only of second-order importance.

A majority of respondents (57 per cent) said their research was essentially open-access while a minority (35 per cent) said it was essentially proprietary (remember though that 41 per cent of respondents belong to the 'Fundamental Research' divisions of the CEA). Moreover, only two respondents answered that their research results were not open-access while only four said their results were not proprietary. In other words, for a large majority of respondents (78 per cent), research results are made open-access *and* appropriated. This suggests that researchers are able to sit on the fence in the sense that they manage to publish some research results (95 per cent) and to patent some others (84 per cent), developing safeguards against the 'secrecy problem'.[8] Here again, though, a weak substitution exists between proprietary and open-access research since the two variables have the same value only in five instances: for 65 per cent of the respondents, research results are appropriated and made open-access with one dominating and one dominated type.

Similar results are found for the 'Technical achievements'/'Fundamental knowledge' dimension. A majority of respondents (57 per cent) see the former as being the essential outcome of their activities while fundamental knowledge is the essential outcome for less respondents (41 per cent). Eighty-one per cent indicate that their activities have both 'technical achievements' *and* 'fundamental knowledge' as an outcome, while there is a weak substitution between the two since one type of output dominates the other in most case (93 per cent of the preceding subset).

Regarding the changes that have occurred in the past few years, the strongest shift in research profiles has been towards more use-inspired research (14 + 54 =68 per cent). Other shifts towards more technical achievements (14 + 32 =46 per cent) and more proprietary research (5 + 46 =51 per cent) have also occurred, yet to a lower extent. When combined, these three trends show that in the aggregate, it is the idealistic 'applied research' type (see subsection 2.3) that has gained increased importance in the past years.

The antagonism between opposite ends of the taxonomy's axes appears in this dynamic (or marginal) perspective since at the same time, 30 per cent of respondents answer that their research is less curiosity-driven, 16 per cent that it is less open-access and 16 per cent that it is less aiming at fundamental knowledge generation. More precisely, I find that use-inspiration and curiosity-drive are strong substitutes at the margin in the sense that both never experience the same shift: when one is said to be stronger, the other is said to be weaker or unchanged (and vice-versa). This strong marginal substitution feature also holds for research aiming at technical achievements vs. research aiming at fundamental knowledge generation. And likewise concerning proprietary vs. open-access research except for one researcher in the Division of Military Applications who declares that her research activities have become much more open-access and at the same time more proprietary, plus another in the Division of Technological Research the activities of whom have simultaneously become more open-access and proprietary.

Interestingly, it should be mentioned that for 27 per cent of respondents, research results are made more open-access. This answer has been made by every researcher but three in the Division of Military Applications, which is overrepresented in this subset (70 per cent). As a matter of fact, a significant part of this Division's programmes have swung from technological to scientific work in the past decade including from the point of view of the norms of disclosure of results, following the French decision to abandon nuclear testing in 1996. Since then, the credibility of French nuclear deterrence heavily relies on the CEA's capacity to demonstrate cutting-edge scientific capabilities for simulation-based nuclear weapon design.

More generally speaking, the survey provides insights of the extent to which the self-governed scientific commons are put in danger at the CEA: for a large majority of respondents, research producing new fundamental knowledge has not diminished in the past few years (84 per cent) nor has open-access research (84 per cent) nor has curiosity-driven research (70 per cent).[9] The latter result is the more pessimistic.

When asked about the reasons for the shift in their research profile over the past few years, one respondent out of two (49 per cent precisely) answer that it is essentially the sources of financial support. It is striking that all these respondents but one consider their research as having drifted towards being more or much more use-inspired in the past few years (subset probability of 94 per cent as compared to 68 per cent in the whole sample). The financial support motivation is also closely associated with the shifts in terms of more or many more technical achievements (subset probability of 78 per cent as compared to 46 per cent in the whole sample) and in terms of more or many more proprietary results (72 per cent vs. 51 per cent). Moreover, all respondents who said that their research had become less or much less open-access are included in these 49 per cent (subset probability of 33 per cent as compared to 16 per cent in the whole sample). The financial support motivation is also closely associated with the shifts towards less or much less curiosity-driven research (50 per cent vs. 30 per cent) and towards less or much less fundamental knowledge generation (28 per cent vs. 16 per cent).

So the story being told by these figures is that of all types of research in the sample being forced to be more use-inspired but also more proprietary and aiming at more technical achievements because of the sources of financial support they have to chase. This shift occurs at the expense of the idealistic fundamental research type (less curiosity-drive, less open-access and less fundamental knowledge generation; see subsection 2.3). These results are summarised in Table 3.4.

It is noteworthy that this story is being told by every researcher in the Division of Technological Research, which heavily depends on non-recurring funding. But incidentally, it is also told by researchers in every division of the CEA (yet only very marginally – i.e. by one researcher – in that of Military Applications where financing comes in a more regular fashion essentially through the Ministry of Defence).

Another 30 per cent consider that their hierarchy is essentially responsible for the shifts that occurred in their research profile. This feature would probably be to a certain extent specific to the CEA where, as compared to its peers in France, hierarchy plays a much greater role. It actually happens that the three technological divisions in the CEA (Division of Military Applications, Division of Nuclear Energy and Division of Technological

Table 3.4 Percentage changes in research profiles when the sources of financial support are an essential drive for change: conditional probability compared to whole sample probability (in brackets)

	Technical achievements	Fundamental knowledge	Use-inspired	Curiosity-driven	Proprietary	Open-access
More + much more	78 (46)		94 (68)		72 (51)	
Less + much less		28 (16)		50 (30)		33 (16)

Research), more hierarchically organised than the Division of Fundamental Research, are over-represented in this subset (subset probability of 82 per cent as compared to 59 per cent in the whole sample). In these divisions, hierarchy is part of the ethos as well as a coordination mechanism of greater importance than in the Division of Fundamental Research – although formal organization charts are similar in every division. Correlatively, these researchers are mostly engaged in activities essentially aiming at technical achievements (82 per cent vs. 57 per cent) and essentially appropriating their results (55 per cent vs. 35 per cent). It is noteworthy that the responsibility put on the hierarchy is primarily associated with lower probabilities to experience no shift in research profiles (except for the shift in terms of technical achievements) but that the shifts contemplated are ambiguous (for example, stronger probability than in the aggregate to have more or much more open-access results but also to have less or much less open-access results).

Lastly, industrial partners are said to be the essential cause of the shifts contemplated in research profiles by another 22 per cent of the respondents. The research profile of these respondents is less oriented towards essentially open-access results (subset probability of 38 per cent as compared to 57 per cent in the whole sample) and more oriented towards essentially proprietary results (50 per cent vs. 35 per cent). The shift under scrutiny is predominantly towards less or much less fundamental knowledge generation (38 per cent vs. 16 per cent). But less or much less curiosity-drive is also in evidence (50 per cent vs. 30 per cent) as well as less or much less open-access results (25 per cent vs. 16 per cent) and more or much more proprietary results (75 per cent vs. 51 per cent). The Division of Technological Research of the CEA that is heavily engaged in collaborative research with industrial partners is overrepresented in this subset (38 per cent as compared to 16 per cent in the whole sample).

5. CONCLUSION

The taxonomy developed in this chapter expands on Stokes's 'Pasteur's Quadrant' and allows better characterisation of research activities and better scrutiny of the changes that they undergo. The survey based on this expanded taxonomy in the French Commissariat à l'Energie Atomique delivers a broad message of hybridisation. Clearly, following Pasteur's Quadrant, it appears that research at the CEA is predominantly 'use-inspired basic research', i.e. it encompasses considerations of use and follows the quest for fundamental understanding; but it is also much more than this 2D taxonomy could ever tell. The survey shows that in the CEA, research activities are by and large both use-inspired and curiosity-driven, both producing fundamental knowledge and technical achievements as well as both made open-access and appropriated. Notice that the basic vs. applied research distinction is practically useless in saying anything about the research that is being conducted there – and, I contend, in many other places.

The survey also reveals that the increased pressure experienced by researchers to take more economic and societal needs into consideration manifests itself essentially through more use-inspiration and more proprietary results. Furthermore, the analysis shows that use-inspired and curiosity-driven research are perfect substitutes at the margin, that is in a dynamic setting. This marginal substitution feature also holds for research aiming at technical achievements vs. that aiming at fundamental knowledge generation, and for research generating proprietary results vs. research producing open-access results.

Tracking sources of financial support for research activities is considered to be the most important factor triggering the changes contemplated. When this cause is said to be essential, the shift in research profiles is even more clearly directed towards more 'pure applied' research (i.e. proprietary and use-inspired research aiming at technical achievements) and less 'pure fundamental' research (i.e. open-access and curiosity-driven research aiming at fundamental knowledge generation). This set of results highlights the impact of the changing rationale for research funding that has been taking place in Europe during the past decades (Geuna 2001) and the importance of the composition of research funding for regulating the research sector overall. It intimates the following main policy implication of this work.

Since Nelson (1959) and Arrow (1962), it is theoretically established that the provision of resources for R&D through the competitive market cannot trigger a sufficient production of scientific knowledge because of its non-appropriability. Conversely, it can be anticipated that the growing contractual-oriented and competitive approach to research funding should prompt more proprietary research – as characterised by the expanded taxonomy – and this is illustrated in the findings. But there is more than that: the results presented above suggest that such competitive funding also skews research towards more directly useful projects (to a greater extent than industrial partners do) and that it stifles the generation of fundamental knowledge.

Moreover, it appeared that researchers are able to perform various or multifaceted activities but that they cannot develop them altogether: the changes that occur underscore the growth of some types of activities at the expense of other types. Accordingly, it is very important for policy-makers to strike a balance between recurring and non-recurring funding for public research according to a choice that should be made explicit in terms of the relative volumes of fundamental and applied research. From this point of view, the consequences of the creation in 2005 of the French 'Agence Nationale de la Recherche' (ANR), a funding agency for research that will increase the volume and proportion of project-based financing, deserve particular attention.

Methodological issues are raised by this study. First of all, the CEA is a government lab and as such, it differs from university research in many respects – in particular, it is presumably much more concerned with economic, industrial and societal needs. Second, the dataset that is used consists in self-reported information about one's research activities, the reliability of which can be tenuous given their important emotional dimension. For example, it may not come as a surprise that not one researcher said his or her activities were not use-inspired at all – an answer that could be emotionally associated with being useless. Last, the results must be considered cautiously since they are based on a survey prototype that covers a sample of very limited size. They will be further consolidated and validated only with the extensive survey that will be carried out shortly.

NOTES

1. Variations of this four-level subdivision have been used among others by Godin (1996); Ranga et al. (2003); Van Looy et al. (2004) and Larsen (2005).
2. The true motivations of researchers may not be that easy to identify. Calvert (2002) shows that in order to compete for funding, researchers may present their activities as use-inspired although their considerations for use are only marginal.
3. Notice that their methodology has close connections with the extension of Pasteur's Quadrant set forth in this chapter.
4. The CEA had 14 910 employees as of December 31, 2006; the CNRS (Centre National de la Recherche Scientifique) ranks first with 26 078 employees.
5. According to a study by OST (2003) on the 1997–2001 period, the CEA was the most actively patenting public research institution in France.
6. The CEA's commitment to provide strong assistance for the design, construction and operation of the International Thermonuclear Experimental Reactor (ITER) is said to

have played a significant role in the decision taken recently by the international community to have it built in France close to one of the CEA's locations in the Provence region (Cadarache research centre).

7. Note that no respondent in the whole sample considers his or her research to be not use-inspired at all; once again, this result is not surprising in the case of the CEA the missions of which are predominantly technological ones.
8. Or to patent results before publishing them; results may also be kept secret in the case of defence-related activities.
9. Note moreover that the changes in terms of fundamental knowledge generation are significantly correlated with those in terms of open-access and of curiosity-driven research (correlation coefficient of +0.7 for both).

REFERENCES

Arrow, K. (1962), 'Economic welfare and allocation of resources for invention', in R. Nelson (ed.), *The Rate and Direction of Inventive Activity*, Princeton, NJ: Princeton University Press, pp. 609–25.

Blumenthal, D., S. Epstein and J. Maxwell (1986), 'Commercializing university research: lessons from the experience of the Wisconsin Alumni Research Foundation', *New England Journal of Medicine*, **314** (25), 1621–6.

Blumenthal, D., E. Campbell, M. Anderson, N. Causino and K. Seashore-Louis (1997), 'Withholding research results in academic life science: evidence from a national survey of faculty', *Journal of the Academic Medical Association*, **277** (15), 1224–8.

Brooks, H. (1994), 'The relationship between science and technology', *Research Policy*, **23**, 477–86.

Bush, V. (1945), *Science, the Endless Frontier. A Report to the President*, Washington, DC: U.S. Government Printing Office.

Calvert, J. (2002), 'Making academic research useful. Scientists' responses to changing policy demands', paper presented to the NPRNet Conference 'Rethinking Science Policy', 21–23 March, SPRU.

Calvert, J. and B.R. Martin (2001), 'Changing conceptions of basic research?', background document for the OECD Workshop on Policy Relevance and Measurement of Basic Research, Oslo, 29–30 October.

Cohen, W., R. Florida and W.R. Goe (1994), 'University-industry research centers in the United States', Carnegie Mellon University report for the Ford Foundation.

Cohen, W., R. Florida, L. Randazzese and J. Walsh (1998), 'Industry and the academy: uneasy partners in the cause of technological advance', in R.G. Noll (ed.), *Challenges to Research Universities*, accessed at http://brookings.nap.edu/books/0815715099/html/ index.html.

Dasgupta, P. and P.A. David (1994), 'Toward a new economics of science', *Research Policy*, **23**, 487–521.

David, P.A. (2000), 'The political economy of science', in H. Lawton Smith (ed.), *The Regulation of Science and Technology*, London: Macmillan.

David, P.A. (2003), 'The economic logic of "Open science" and the balance between private property rights and the public domain in scientific data and information: a primer', SIEPR discussion paper no. 02-30.

Florida, R. (1999), 'The role of the university: leveraging talent, not technology', *Issues in Science and Technology* (Summer), 363–73.

Florida, R. and W.M. Cohen (1999), 'Engine or infrastructure? The university role in economic development', in L.M. Branscomb, F. Kodama and R. Florida (eds), *Industrializing Knowledge. University–Industry Linkages in Japan and the United States*, Cambridge, MA: MIT Press.

Geuna, A. (2001), 'The changing rationale for European university research funding: are there negative unintended consequences?', *Journal of Economic Issues*, **35** (3), 607–32.

Gibbons, M. (1999), 'Science's new social contract with society', *Nature*, **402**, 81–4.

Goddard, J.G. and M. Isabelle (2006a), 'How do public laboratories collaborate with industry? New survey evidence from France', IMRI working paper no. 06/02, June.

Goddard, J.G. and M. Isabelle (2006b), 'Managing intellectual assets within knowledge-based partnerships: insights from a survey of public laboratories collaborating with industry', IMRI working paper no. 06/03, July.

Godin, B. (1996), 'Research and the practice of publication in industries', *Research Policy*, **25**, 587–606.

Heller, M. and R. Eisenberg (1998), 'Can patents deter innovation? The anticommons in biomedical research', *Science*, **280**, 698–701.

Henderson, R., A.B. Jaffe and M. Trajtenberg (1998), 'Universities as a source of commercial technology: a detailed analysis of university patenting, 1965–1988', *Review of Economics and Statistics*, **80**, 119–27.

Hicks, D. and K. Hamilton (1999), 'Does university-industry collaboration adversely affect university research?', *Issues in Science and Technology*, **15** (4), 74–5.

Larsen, M.T. (2005), 'Does industrial relevance in public science come at the expense of basic research?', paper presented at the DRUID Winter Conference, Aalborg/Rebild Bakker, 27–29 January.

Merton, R.K. (1973), *The Sociology of Science: Theoretical and Empirical Investigations*, Chicago, IL: University of Chicago Press.

Mowery, D.C., R.R. Nelson, B.N. Sampat and A.A. Ziedonis (2001), 'The growth of patenting and licensing by U.S. universities: an assessment of the effects of the Bayh–Dole Act of 1980', *Research Policy*, **30**, 99–119.

Murray, F. and S. Stern (2006), 'Do formal intellectual property rights hinder the free flow of scientific knowledge? An empirical test of the anti-commons hypothesis', paper presented at the Druid Summer Conference, Copenhagen, 18–20 June.

Narin, F., G. Pinski and H.H. Gee (1976), 'Structure of the biomedical literature', *Journal of the American Society for Information Science*, **27**, 25–45.

Nelson, R.R. (1959), 'The simple economics of basic scientific research', *Journal of Political Economy*, **67** (3), 297–306.

Nelson, R.R. (2000), 'Knowledge and innovation systems', in OECD (ed.), *Knowledge Management in the Learning Society*, Paris: OECD.

Nelson, R.R. (2004), 'The market economy and the scientific commons', *Research Policy*, **33**, 455–71.

OECD (1994), *Main Definitions and Conventions for the Measurement of Research and Experimental Development (R&D). A Summary of the Frascati Manual*, Paris: OECD.

OECD (1996), *The Knowledge-based Economy*, Paris: OECD.

OST (2003), *Rapport sur les indicateurs relatifs à la propriété intellectuelle dans les organismes de recherche publique et dans les établissements d'enseignement supérieur*, Paris: OST.

Pavitt, K. (2001), 'Public policies to support basic research: what can the rest of the world learn from US theory and practice? (and what they should not learn)', *Industrial and Corporate Change*, **10** (3), 761–79.

Polanyi, M. (1962), 'The republic of science: its political and economic theory', *Minerva*, **1** (l), 54–73.

Ranga, L.M., K. Debackere and N. von Tunzelmann (2003), 'Entrepreneurial universities and the dynamics of academic knowledge production: a case study of basic vs. applied research in Belgium', *Scientometrics*, **58** (2), 301–20.

Scotchmer, S. (1991), 'Standing on the shoulders of giants: cumulative research and the patent law', *Journal of Economic Perspectives*, **5**, 29–41.

Shapiro, C. (2001), 'Navigating the patent thicket: cross licenses, patent pools, and standard-setting', in A. Jaffe, J. Lerner and S. Stern (eds), *Innovation Policy and the Economy*, Cambridge, MA: MIT Press.

Stokes, D.E. (1997), *Pasteur's Quadrant: Basic Science and Technological Innovation*, Washington, DC: Brookings Institution Press.

Van Looy, B., M. Ranga, J. Callaert, K. Debackere and E. Zimmermann (2004), 'Combining entrepreneurial and scientific performance in academia: towards a compounded and reciprocal Matthew-effect?', *Research Policy*, **33**, 425–41.

Walsh, J.P. and W. Hong (2003), 'Secrecy is increasing in step with competition,' *Nature*, **422**, 801–02.

Walsh, J.P., C. Cho and W.M. Cohen (2005), 'Patents, material transfers and access to research inputs in biomedical research', final report to the National Academy of Sciences' Committee on Intellectual Property Rights in Genomic and Protein-Related Inventions, September.

Ziman, J. (1994), *Prometheus Bound: Science in a Dynamic Steady State*, Cambridge: Cambridge University Press.

APPENDIX 3.1

Questionnaire Used for the Survey

Q0 *Are you pleased with this taxonomy?*

❑ Yes ❑ More or less ❑ No

Why? type your comments here . . .

Q1 *Please position your research activities in this taxonomy:*

	Yes essentially	Yes to a lower extent	No
Technical achievements	❑	❑	❑
Fundamental knowledge	❑	❑	❑
Use-inspired	❑	❑	❑
Curiosity-driven	❑	❑	❑
Open-access results	❑	❑	❑
Proprietary results	❑	❑	❑

Q2 *Has this position changed in the past few years?*

	Much more	More	Unchanged	Less	Much less
Technical achievements	❑	❑	❑	❑	❑
Fundamental knowledge	❑	❑	❑	❑	❑
Use-inspired	❑	❑	❑	❑	❑
Curiosity-driven	❑	❑	❑	❑	❑
Open-access results	❑	❑	❑	❑	❑
Proprietary results	❑	❑	❑	❑	❑

Q3 Have these changes (if any) operated under the pressure of:

	Yes essentially	Yes to a lower extent	No
Industrial partners	❑	❑	❑
Sources of financial support	❑	❑	❑
Evaluation systems	❑	❑	❑
Hierarchy	❑	❑	❑
Other (please develop . . .)	❑	❑	❑

Q4 Additional information:

Division: ☐

Research fields: Field 1 Field 2 Field 3 Field 4

(use the following link to the Dewey Decimal Classification Index: http://www.scit.wlv.ac.uk/wwlib/browse.html)

APPENDIX 3.2: BAR CHARTS OF RESPONSES

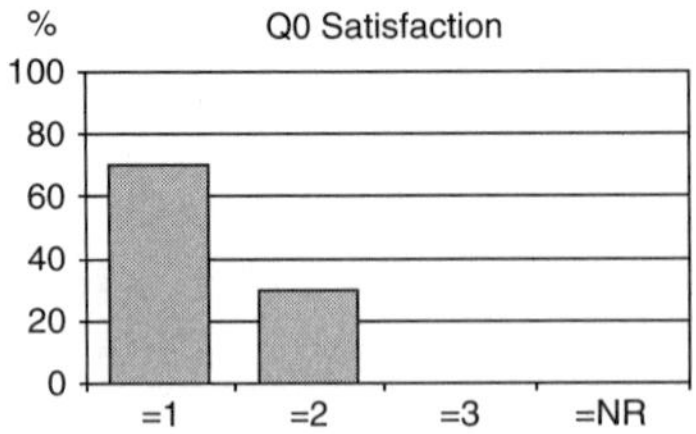

Key

1 yes 2 more or less 3 no (Q0)
1 yes essentially 2 yes to a lower extent 3 no (Q1)
1 much more 2 more 3 unchanged 4 less 5 much less (Q2)
1 yes essentially 2 yes to a lower extent 3 no (Q3)

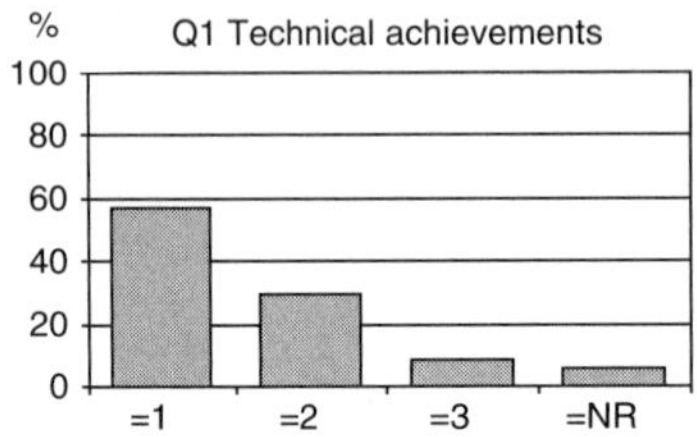

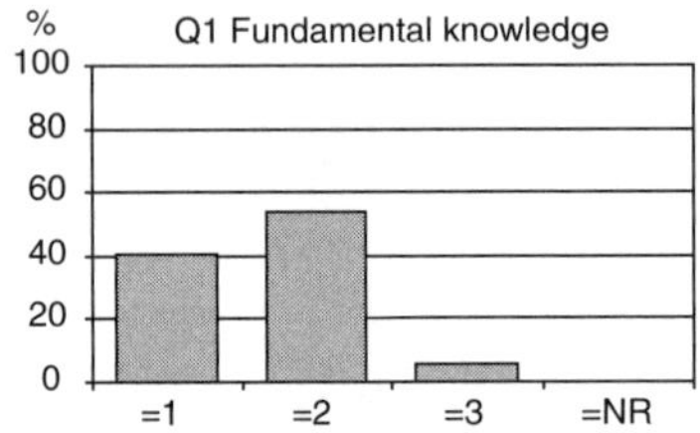

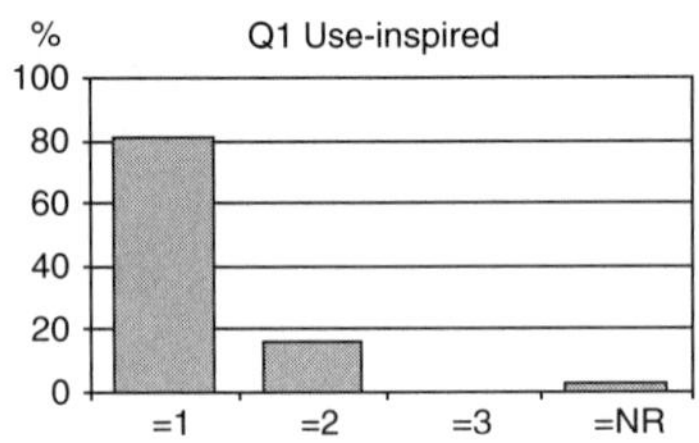

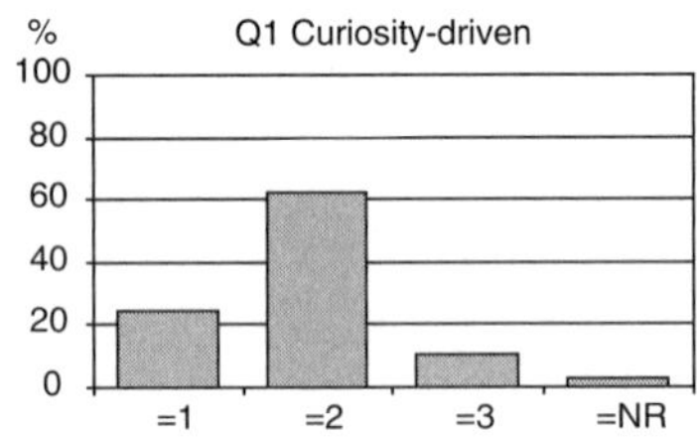

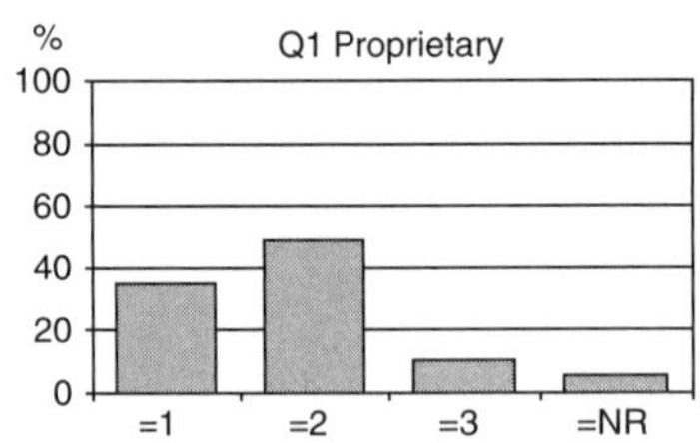

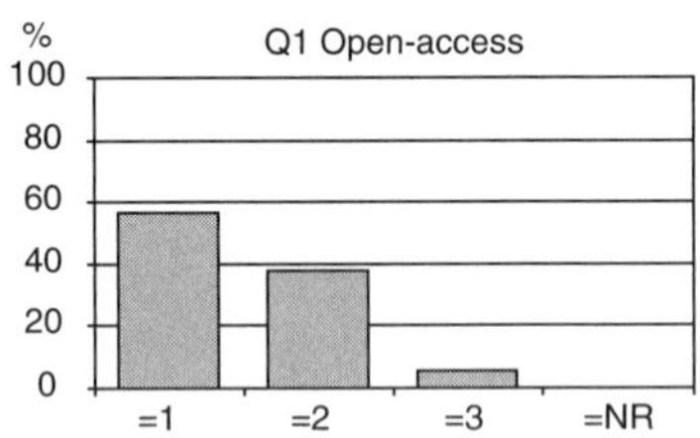

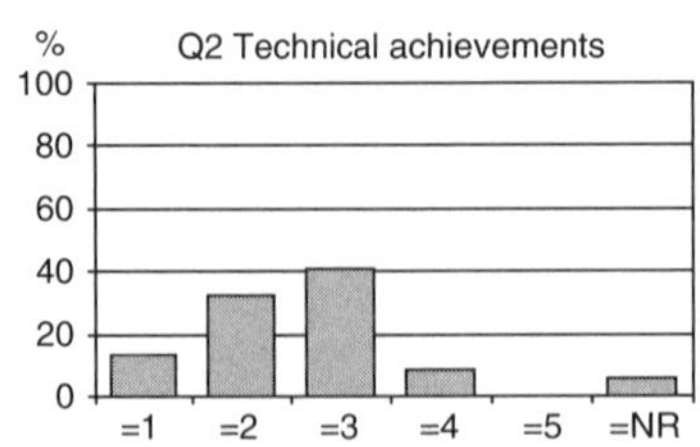

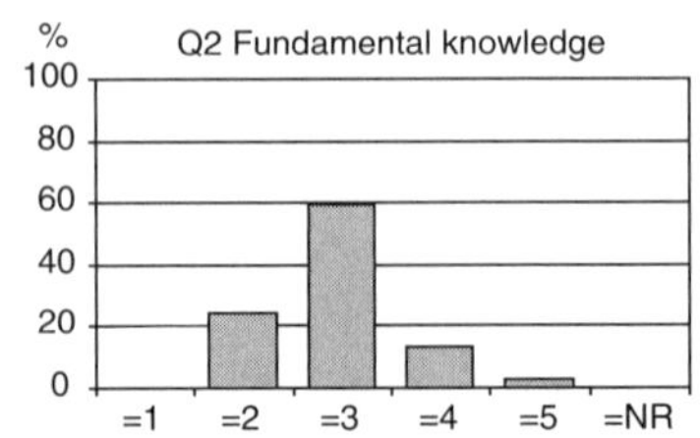

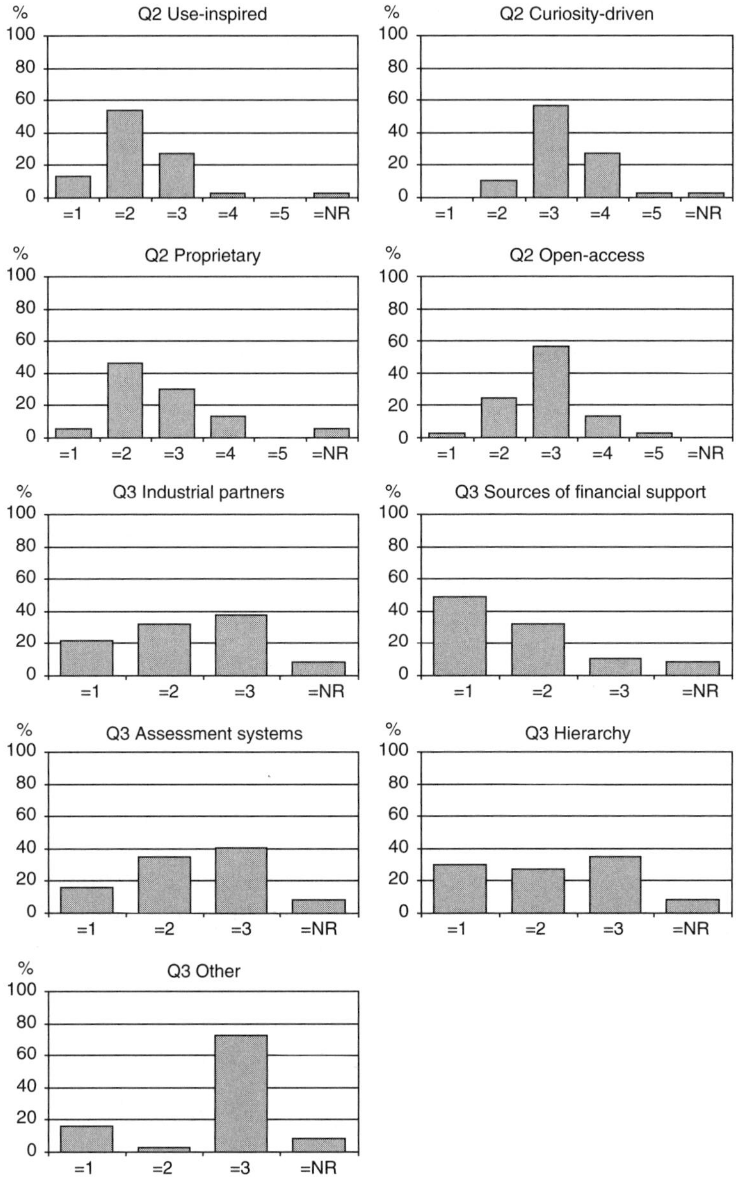
Q2 Use-inspired
Q2 Curiosity-driven
Q2 Proprietary
Q2 Open-access
Q3 Industrial partners
Q3 Sources of financial support
Q3 Assessment systems
Q3 Hierarchy
Q3 Other
%
100
80
60
40
20
0
=1
=2
=3
=4
=5
=NR

4. Towards an integrated patent system and innovation prospects in Europe

Alfredo Ilardi and Blandine Laperche

Patenting is as an essential tool for the genesis of innovation, giving incentives to investments by enterprises and stimulating the creativity of a society. 'The patent system added the fuel of interest to the fire of genius,' declared Abraham Lincoln.[1] Since its origin, it has been an entitlement to property that is spatially well delimited, but the creation of specific patent systems in each country was soon considered as a barrier to trade. By patent systems we mean, referring to Lévêque and Ménière (2006, p. 11), 'the set of institutions allowing the application of patent laws in a given geographic area'. The harmonization of patent laws (and more globally of intellectual property rights) has thus been a recurrent question since the end of the 19th century (Ilardi 2005). In this chapter, we study the harmonization of patent laws in Europe, which includes the foreseen creation of an integrated patent: the 'Community patent'.

How can this willingness to harmonize the rules dealing with the filing and grants of patents to create a unique patent in Europe be explained? What has been the institutional history of the emergence of a unified patent law in Europe? We will argue that this history has been and is still paved with difficulties and is the subject of hard negotiations, which show not only the economic but also the political character of the property resulting from inventions. Finally, we defend the idea that the reforms of the patent system in Europe should be integrated into scientific and technological policies and that these, except in political declarations, do not seem to have sufficient priority within European economic policy to lead to positive results in terms of innovation.

1. HOW CAN THE EMERGENCE OF AN INTEGRATED PATENT SYSTEM IN EUROPE BE EXPLAINED?

1.1 Roles and Characteristics of Patents

It is nowadays widely acceptable to consider the invention patent as a major incentive tool for innovation and for the disclosure of scientific and technical information. The patent gives incentives to innovators because the exclusive right granted allows the patent holder to recoup his/her investment. It favours the dissemination of scientific and technical information, thus ensuring the cumulative character of the innovation process. On the one hand, the exclusive right is temporary and at its end the invention passes into the public domain and may be then used by everyone. On the other hand, the patent holder has to publish a description of his/her invention. The issued description must be sufficiently accurate that a specialist would be able to reproduce it (Lévêque and Ménière 2003; Scotchmer 2004).

However, to understand the emergence and the stakes linked to the development of an integrated patent system in a regional area, and as far as we are concerned here in Europe, it seems important to stress two fundamental and linked characteristics of patents: first, since its origin, the patent has been a title of property conferred by the states. Second, its legal evolution and the economic strategies which are associated with it show that patents symbolize the well-known adage according to which 'knowledge is power' ('Scientia potentia est', Sir Francis Bacon). The first privileges that share some characteristics with today's patents (exclusive right of use, disclosure of information) appear with the building of European states. The latter were first of all cities, as was the case of Venice, which promulgated in 1474 the first law protecting industrial property. The first privileges conferred on inventors aimed to attract the most improved techniques. Scientific and technical knowledge and its materialization in invention were considered as tools of economic and political power. The English Statute of Monopolies, dated 1624, gave the possibility of protecting an imported technique. In France, the royal privileges of the 17th century were integrated into state policy for development, influenced by mercantilist ideas, which notably in France, remained reliable to Colbert's ideas of attracting the best technicians. The French Revolution put an end to privileges but the Republic placed Property at the first rank of human rights. The French Patent Law of the 7 January 1791 protected national inventions and imported patents on national territory. The economic and political competition between states largely explains this fact as France, lagging behind Great Britain which was beginning its industrial revolution, was trying to catch up (Beltran et al. 2001; Hilaire Perez 2000; Laperche 2001).

Subsequently patent were limited to real inventions and no longer for imported techniques (the 'importation patent' was for example suppressed in France by the revision of 1844), yet the link between appropriation of inventions and economic power is still very important. For example, despite its limits, patents are still the key indicator to compare the capacity to invent of firms and countries (OECD 2002). Moreover, the recent evolution of intellectual property rights, with the widening of patentability to new fields at the frontier of scientific discovery and invention (notably in the field of biotechnology) (Gallini 2002), show that appropriation (and thus patenting) is one of the elements of the strategic arsenal of enterprises. They use it as a defensive means of protection against imitations but also as an offensive tool, to create new markets, to define and impose standards and also to lure competitors (Laperche 1998). Patent applications for biological 'inventions', ensuing from traditional knowledge transmitted from generation to generation, show that the notion of novelty is sometimes understood in a restrictive way (within a geographic area, or according to a mode of expression e.g. written versus oral, Mgbeoji 2005), according to the interest of states and their national champions (sees the debates on this question at http://www.wto.org).

To summarize, since its original conception, patenting has been a territorial right, which means that a patent delivered in a country is in force only in that country. This territorial right, which has evolved to take its contemporary characteristics, has been used and is still used to affirm the industrial power of a country and of its firms compared to their competitors. However, efforts at harmonization started at the end of the 19th century and have continued during the 20th century and beyond.

1.2 Harmonization and Globalization

Several reasons may explain the attempts to harmonize national laws on industrial property and to create integrated patent systems at the regional and international level: the globalization of economies and the affirmation of regional poles in the global economic area, are among the major reasons.

It is possible to distinguish three main steps in the harmonization of industrial property rights at the international and regional scales (we focus here on Europe), which correspond to important phases in the globalization of economies.

The first step took place in the 19th century when the first international conventions began, like the Paris Convention (1883) for the protection of industrial property and the Berne Convention (1886) for the protection of literary and artistic works. The national particularities of industrial property laws were considered as barriers to trade, and this approach gained ground at the end of the 19th century, notably thanks to the development of means communication (railways, telegraph). In this period, the numerous universal exhibitions aroused the apprehensions of exhibitors about the possibility of copying the exhibited techniques and stimulated international negotiations. It is to be noted that whereas the term 'globalization' is a concept used to characterize today's economy, it is however an old notion. The first phase of globalization was based on commercial trade, and was rooted in the constitution of nation-states and the discovery of new territories at the end of the 15th century. At the end of the 19th century a second step was taken towards globalization, where not only trade but also production developed at an international scale (it was at that time that the first multinational enterprises were born). The end of the 19th century was also characterized by an important liberalization of markets and of financial flows between countries, which were then partitioned off, even more strongly after World War I (Bairoch 1993).

The second period in the harmonization process of intellectual property rights began after World War II and ran to the end of the 1970s. During this period the international system of patenting (the Patent Cooperation Treaty, PCT)[2] was created as an answer to the growing number of patent applications coming from abroad, and also the World Intellectual Property Organization (WIPO, established in 1967), today a specialized agency of the United Nations dedicated to the promotion of intellectual property throughout the world and to the administration of intellectual property treaties. It was also during this period that the European patent system (European Patent Convention – EPC)[3] was established, whose history is detailed in the second part of this chapter. The Bretton Woods institutions (the World Bank and the International Monetary Fund) were created in the aftermath of World War II with the aim to reducing the barriers to trade. The role of the General Agreement on Tariffs and Trade (GATT), signed in 1947, was to reduce tariff barriers between countries. The trauma of World War II led to the creation of common interests in Europe, to avoid conflicts. The European Coal and Steel Community in 1951, and the Treaty Establishing the European Economic Community (Treaty of Rome) signed in 1957 built up a European economic area with the aim of promoting economic and political integration among member states: the creation of a free trade area would justify the reduction of non-tariff barriers to trade, as is the case of intellectual property.

The third period began in the 1980s and is characterized by the globalization of intellectual property rights, which culminated with the conclusion in

1994 of the Agreement on Trade Related Aspects of Intellectual Property Rights (TRIPS agreement). This unprecedented harmonization of intellectual property rights in all the member states of the World Trade Organization (WTO, which replaced GATT in 1995) corresponds to the third step in the globalization process, into which trade, production and finance are simultaneously integrated. In fact, the deregulation of financial markets in the 1980s has created a large interconnected market, and finance thus has become the leading point of globalization (Chesnais 1996). Global firms are networked firms: their conception, production and commercial processes are directly organized on a global scale (Castells 2001; Andreff 2003; Chesnais 1996; Laperche et al. 2006). In this context, the globalization of intellectual property rights has become a necessity to secure their investments and their technology transfers (Maskus and Reichman 2004).

The current stage of globalization, in which all three spheres of the economy are involved (trade, production, finance) is an important explanation of the need to harmonize the rules governing industrial property. The regional character of globalization adds an explanation of the importance of integrated patent systems, notably in Europe. The Triadic character of globalization, according to Ohmae's work (1985), implies that economic activities (trade, material, intangible and financial investments) are largely concentrated in three zones: the United States of America, Europe and Japan, although these are currently broadening. Looking at merchandise trade: in 2005, 73.2 per cent of Europèan total merchandise exports (in dollars) were intra-regional and 31.5 per cent of world merchandise trade flows were concentrated in the European zone (WTO 2006). The significant role played by the Triad in the economic activities – and within the Triad, of Europe – can be seen in foreign direct investment (FDI) statistics. Over the past decades, the share of the Triad in total world inward FDI flows and stocks has fluctuated at around 60-70 per cent. The share of European Union in FDI inflows into the Triad was 75 per cent in 2003-05, compared to 62 per cent in 1978–90. In 2005, the European Union (including the eight economies formerly classified under Central and Eastern Europe), accounted for almost half of global inward and outward flows and stocks (UNCTAD 2006). The intensity of economic relations gives incentives to the creation of common institutions, notably in the field of intellectual property. In this line of thought the resumed discussions on the Community patent in the 1990s go hand-in-hand with economic integration: 'One market, one currency, one patent', declared R. Goebbels, Minister of the Economy for the Grand Duchy of Luxembourg, at the opening of the hearing on the Green Paper on patents in Europe, Luxembourg, 25 November 1997 (Schmiemann 1998).

1.3 Harmonization, Globalization and Innovation

Since the end of the 1970s and the questioning of the Fordist mode of production, innovation – defined by J.A. Schumpeter ([1911] 2006 and [1942] 1976) as the introduction of new combinations of productive resources (new product, new process of production, new organization of production, new market, use of a new source of raw material) – has become a major element in economic competition (Porter 1998) and of the economic growth of nations (as in theories of endogenous growth, see Aghion and Howitt 1998). In contemporary economics, it is acknowledged that innovation needs a certain degree of monopoly, hence it needs patent protection (Schumpeter [1942] 1976; Laperche 2004), notably because of the particular characteristics of knowledge (partly tacit, non-rival, non-excludable and cumulative) which reduce the private return of its investment (Arrow 1962; Nelson 1959). Patenting is a key device to increase the private return on investment in knowledge, despite its limitations, even if it is not the only element useful for the promotion of knowledge-based economies (public investment in basic research, development of internal learning to create routines and to limit externalities of knowledge are some important devices to enhance the innovation capacities of firms and the appropriation by the firm of the knowledge it has developed).

Establishing the role of innovation in economic growth (and thus the limits of Solow's model of growth (1957) where technical progress is exogenous) has led in economic analysis to the development of studies aiming at a better understanding of the 'black box' of technology (Rosenberg 1982) and of the institutional forms involved in the genesis and the dissemination of innovation. The economists specialized in the field of technical change have worked since the 1980s on the nature and the role of institutions that, within the national framework, can promote innovation (Freeman 1987; Lundvall 1992; Edquist 1997). However, in the context of globalization, is the national level still pertinent to study the genesis and the dissemination of innovation? On the one hand, the answer is affirmative because the diversity of institutional arrangements in the field of innovation provides an answer to the relative strength of nations and enterprises which compete at the global level (Amable et al. 1997). On the other hand, with the harmonization/unification of some regional areas, as is the case in Europe, views have developed on the emergence of a European system of innovation, made up of 'post-national' institutions dedicated to innovation (Caracostas and Soete 1997).

However, one of the major elements of the functioning of an innovation system is the legal framework of accumulation, that is the set of rules and laws framing the economic activities, including innovation (Uzunidis 2003).

The legal framework of accumulation is built by state intervention. Three main policy orientations were proposed, first in the USA, to obtain improved results in the field of innovation (Branscomb and Keller 1998): (1) financing activities individually or collectively generating appropriable resources; (2) creating devices allowing the re-appropriation of the private return on R&D investments; (3) setting up cooperation procedures between public and private entities in order to ensure the profitability of private investments with strong positive external effects. These policy orientations reflected a move from mission-oriented programmes to distributed science and diffusion-oriented programmes. These aims have also been diffused throughout the world, notably in Europe (von Tunzelmann and Nassehi 2004). Concerning European intervention in the field of innovation, the first aspect has been achieved mainly through the framework programmes, together with other financing programmes. The third policy orientation has also received great attention through the policy of network building (clusters, commercialization of public research, innovation collaboration). However the second aspect (creation of devices allowing the re-appropriation of the private return of investment in R&D) even if presented as a major objective, has still to be improved. In fact, the measures pointed to by European policy to increase patenting activity for European enterprises are the encouragement of SMEs to apply for patents, the setting up of programmes to disseminate patent information, and policies to encourage public sector research institutions to apply for patents (Arundel and Hollanders 2005). The fragmented character of the European system of patents is not mentioned. This aspect was however particularly stressed by the White Paper on Innovation, dated 1995, which noted the existence of common institutions dedicated to the stimulation of innovation in Europe and also noted their fragmentation. One of the recommendations of this report was thus to promote harmonization, especially in the field of intellectual property. We can link this recommendation with the revival of the Community patent in the 1990s (see section 2).

Knowledge-based economies are characterized by the necessity for firms to constantly renew their supply to meet a changing demand and also to contribute to the creation of new markets. This 'permanent innovation' (Foray 2004) means that the cost of innovation is on the increase, as can be observed in the growing business expenses dedicated to R&D: among OECD countries, they grew by about 50 per cent between the early 1990s and early 2000s, while at the same time GDP increased about 25 per cent (OECD 2004). As a consequence, the management of innovation has been largely transformed. The innovation strategies of large enterprises do not only rely on the gathering of internal resources (human, scientific and technical, financial) but also on collaborations (the signing of contracts, including licensing) with other (small or large) enterprises and other institutions (universities, research centres) (Tidd et al. 2005; Antonelli 2005). The result is a 'socialization of the formation of the enterprises' knowledge capital' (Laperche 2007), which means that the firm's knowledge capital (the set of resources gathered, analysed and used in the production process with the aim to innovate) is not built by a single enterprise but through a network of enterprises and institutions in close cooperation. These collaborations may be achieved at a global level but a large number of studies show that an important part of R&D investments of European multinational corporations take place in the European area (Cantwell and Iammarino 2003). This is the same for R&D partnerships that usually have a regional basis. Hagedoorn (2002) showed that during the four last decades of the 20th century, about 99 per cent of R&D partnerships between companies took place within the Triad. However, his study also showed that intra-North American R&D partnerships were more numerous than intra-European ones and that the trend for intra-European partnerships was gradually eroding. The reasons may be found in the leadership of North America in science and technology and the attractiveness of this market. But it may also be explained, as far as Europe is concerned, by the fragmented institutional framework. However, this geographically designed socialization of the innovation process is another justification for the harmonization of intellectual property rights and the creation of an integrated system of patents in Europe.

2. HISTORY OF THE CONSTRUCTION OF AN INTEGRATED PATENT SYSTEM IN EUROPE

2.1 Early Attempts of International Harmonization of Certain Aspects of Patent Laws

The first three years of the 1970s were decisive in determining the present scenario of the international legal protection of inventions in general and in particular for the establishment of a European patent system.

On 9 June 1970 the Patent Cooperation Treaty (PCT) was signed in Washington and three years later, on 5 October 1973, the diplomatic conference for the adoption of the Convention for the Grant of European Patents (European Patent Convention) (EPC) concluded its deliberations in Munich with the signing of the Convention.

Both of these international instruments, even if different as to their respective legal nature and territorial scope, had their roots in the various attempts to internationalize different aspects of the industrial property system which took place at the end of the 19th century and successfully continued into the 20th century.

The field of trademarks was the first in which international harmonization efforts yielded positive results. On 14 April 1891 the conclusion of the Madrid Agreement Concerning the International Registration of Marks set up an international trademark registration system which is still in force today. In the field of patents several factors delayed the achievement of a similar result until after the conclusion of World War II, even if the debate was gaining ground. At the turn of the 19th century, there was a general reluctance among countries that were members of the Paris Convention for the Protection of Industrial Property to depart from their well-established individual legal rules for the protection of inventions and enter into multilateral negotiations for the harmonization of their national patent systems. This attitude was enhanced by the economic crisis between the two world wars and by its consequences for industrial development, which encouraged competent authorities, supported by interested circles, to maintain the 'status quo' as regards national patent laws.

After the end of World War II, following the general restarting of industrial and commercial activities during the 'reconstruction' period, the attitude of national patent offices and of industry started to change. On the one hand, national administrations faced a substantial increase in the burden of administering their patent systems due to the growth in the number of patent applications filed; on the other hand, the users of the system experienced difficulties in finding adequate responses to their needs in the existing patent legislations.

This change of attitude manifested itself mainly in Europe, thus opening the way to the first attempt at setting up of a European system for the protection of inventions. On 6 June 1945 the Agreement for the establishment at The Hague of the International Patent Institute, with the task of carrying out novelty searches in the field of patents, was signed among a number of European countries. This Agreement was followed, under the auspices of the Council of Europe, on 11 December 1953 by the Convention Relating to the Formalities Required for Patent Applications and on 27 November 1973 by the Convention for the Unification of Certain Points of Substantive Law on Patents for Inventions.

The 1962 Draft Convention on the Establishment of a Community Patent System

However, it was only in the early 1960s that the first concrete European attempt in this direction took place in the framework of the 1957 Treaty Establishing the European Economic Community (EEC). The EEC Treaty does not contain common provisions on the protection of industrial property. It leaves member states free to adopt their own legislation in this field. In 1962, with a view to fostering the free movement of goods within the internal market, thus ensuring that competition is not distorted, a draft convention for the establishment of a patent system providing for the grant of uniform protection to inventions throughout the Community was prepared. But in the early 1960s, the European integration process was just taking its first steps and the achievement of such an ambitious legislative harmonization project soon appeared unrealistic. In 1965, the preparatory work for the preparation of the Convention was interrupted due to political differences among the negotiating states.

The 1969 initiative of the EEC Council on the EPC and CPC

A decade later, as a consequence of new developments in the commercial relations among European countries and of the increase in the number of patent applications filed in those countries, the debate on the need for setting up a European patent system was resumed. In 1969, the Council of the European Communities endorsed the main guidelines which ought to govern the building of such a system. Two international instruments were foreseen: one with a broader territorial scope that would have been applicable to a larger number of European states; the other would have covered only the Community member states.

The first of the two instruments became the European Patent Convention (EPC), the second the Convention for the European Patent for the Common Market (Community Patent Convention) (CPC), which was adopted at Luxembourg on 15 December 1975 and was amended on 15 December 1989 by the Agreement Relating to Community Patents.

The EPC was adopted in Munich in 1973 and entered into force on 7 October 1977, while the CPC never entered into force due to the lack of necessary ratifications. The EPC, which was last revised on 13 December, 2007, created a European 'regional' system for the protection of inventions. It applied as of January 2008 to 34 European states, including the member states of the European Community, which gradually joined the Convention.[4] It is administered by the European Patent Organization, with its headquarters in Munich. The Convention provides for a single procedure for the grant of the 'European patent', which is valid for a period of 20 years and has the same effect as a national patent in each of the member states of the EPC. The processing of applications for the European patent, which includes an examination and an opposition procedure, is the task of the European Patent Office (EPO). Together with the Administrative Council, the EPO is one of the two constituent bodies of the European Patent Organization. The EPC provides a link with the PCT, so that applicants for a European patent may also obtain protection for their inventions in the member states of the PCT, via the PCT procedure.

The CPC[5] was intended to transform the different national patent procedures of the member states of the European Community into one common procedure through which a single patent (the 'Community patent') would be granted. The CPC was implanted in the EPC and through the procedure provided in the EPC opened the way, upon designation of one or more member states of the European Community, to the grant of a supranational title for the protection of inventions, having effect in the whole territory of the Community. To this end, the CPC established within the EPO a number of special departments responsible for the acts of the EPO in respect of the procedure for the grant of the Community patent: the Patent Administration Division, the Revocation Division and the Revocation Board. Furthermore, with the CPC directly linked to obtaining the objectives of the EEC Treaty, the jurisdiction over the interpretation of the Convention was conferred on the Court of Justice of the European Community.[6]

The non-entry into force of the CPC, mainly due to the high cost of the Community patent, in particular that of translation, and the highly complex judicial system foreseen, meant that the other Convention – the European Patent Convention (EPC) – designed with the CPC to set up a European patent system, was left as the only means to regulate the protection of inventions at European level. Indeed, since the date of its entry into force in 1977, the EPC has centralized – and continues to centralize – the procedure for the grant of patents for its member states and has proved to be a major factor in harmonizing the European national patent laws and in fostering the European economic integration process.

2.2 New Attempts in the 1990s to set up a Community Patent System

However, notwithstanding the prominent role played by the EPC in forging European patent law, the idea of establishing an autonomous patent system common to the member states of the EEC was never completely abandoned. From the beginning of the 1990s, the debate on this project was kept alive among competent national administrations and 'interested circles', especially by the Commission of the EEC. But it was only towards the end of the decade, in 1997, with the Commission's Green Paper on the Community Patent and the European Patent System[7] that the initiative of setting up a Community patent was concretely launched. The Green Paper was followed in 1998 by a series of consultations which involved different bodies[8] of the Community and culminated on 5 February 1999 with the adoption by the Commission of a Communication[9] on the Green Paper. In essence, the Communication contained the main elements of the structure of a future Community patent system. On 24 March 2000, the European Council meeting in Lisbon officially endorsed the creation of the Community patent system.

The Commission's Proposal for a Council Regulation on the Community Patent (2000)

The decision taken by the European Council at the Lisbon meeting was followed on 1 August 2000 by a Proposal for a Council Regulation on the Community Patent,[10] presented by the Commission. This proposal, as amended by successive modifications, still constitutes today the platform on which the negotiations for the establishment of the Community patent system are based.

The declared aim of the Council Regulation is the creation of a supranational patent right applicable within the territory of the Community. The unitary character of this right is however conceived to coexist, on the one hand, with national patent rights, and on the other with the European patent right as it stems from the EPC. This means that with the entry into force of the Community patent, patentees would have the choice between three alternatives when seeking protection for their inventions: the national patent, the European patent under the EPC and the Community patent.

Together with the existing Council Regulations on trademarks[11] and on industrial designs,[12] the Regulation on the Community Patent is intended to complete the building of the Community industrial property system with a view to further enhancing the free movement of goods within the Community.

The Proposal for a Council Regulation is composed of a Preamble and six Chapters:

- Chapter I – General Provisions;
- Chapter II – Patent Law;
- Chapter III – Renewal, Lapse and Invalidity of the Community Patent;
- Chapter IV – Jurisdiction and Procedure in Legal Actions relating to the Community Patent;
- Chapter V – Impact of National Law;
- Chapter VI – Final Provisions.

The Preamble deals with the relationship between the future Community patent system and the European patent system under the EPC. This is a focal point in the creation of the Community patent system. The relationship has a particular significance in view of the interconnections between the two systems which imply, once the Regulation on the Community patent were to be adopted, ad hoc revisions of the EPC in order to render the two instruments 'compatible'. In essence, the main points covered by the Preamble are: the performance by the European Patent Office of the function of granting and administering the Community patent; the accession of the Community to the EPC and the creation of a 'Community Intellectual Property Court', which should have jurisdiction on matters relating to the infringement and validity of the Community patent, while the Commission's decisions should be subject of appeal before the Court of Justice of the European Community.

Chapter I (General Provisions) establishes the Community patent law and sets the principle that it applies to the patents granted as Community patents by the European Patent Office. Furthermore, it specifies that the Community patent has a unitary character in the sense that it shall be valid throughout the territory of the Community, and shall have an autonomous character in the sense that it shall be subject to the provisions of the Regulation.

Chapter II (Patent Law) contains a set of substantive law provisions dealing with the rules governing the acquisition of the right to the Community patent, including the relationship between the employee and the employer, the scope of the rights conferred by the Community patent and patent application, and the situations involving the Community patent as an object of property, such as transfer, contractual licences, licences of rights and compulsory licences. It is to be noted that most of these provisions incorporate the content of the corresponding applicable provisions of the CPC, in the EPC and in the Agreement on the Trade Related Aspects of Intellectual Property Rights (TRIPS Agreement 1994).

Chapter III (Renewal, Lapse and Invalidity of the Community Patent) deals on the one hand with the procedures governing the renewal, surrender and lapse of the Community patent and, on the other, with the grounds and effects of invalidity of the patent.

Chapter IV (Jurisdiction and Procedure in Legal Actions Relating to the Community Patent) covers three procedural aspects when the Community patent is subject to actions. The first relates to actions concerning the validity, infringement and use of the Community patent. The Community Intellectual Property Court has jurisdiction in these actions and appeals may be lodged against decisions of the Court to its Chamber of Appeals. The second concerns actions which do not fall within the jurisdiction of the Community Intellectual Property Court, such as proceedings relating to the right to the patent between employer and employee. In such cases, national courts of member states have jurisdiction. The third covers arbitration proceedings involving the Community patent, with the restriction however that they cannot invalidate a Community patent.

Chapter V (Impact of National Law) sets out the prohibition of dual patent protection for the same invention by a national patent of a member state and the Community patent. The same applies to utility models and utility certificates, where such title is provided for by the national law of a member state.

Chapter VI (Final Provisions) contains provisions relating to the establishment of the Register of Community Patents, the publication the *Community Patent Bulletin* and the adoption of the implementing regulations.

The Common Political Approach (2003)

Since its presentation in August 2000, the Proposal for a Council Regulation on the Community Patent has been the subject of intense negotiations, resulting in a large number of working documents[13] containing comments and amendments to the original proposal. It was soon evident that the main subjects requiring a compromise solution were those which had a direct bearing on the activities of the national administrations of the member states interested in the enforcement of the Regulation. They concerned the structure of the jurisdictional system of the Community Patent, the language regime applicable to the Community patent applications and patents, the costs for the maintenance of the Community Patent, the role of national patent offices in the administration of the Community Patent system and the distribution of fees among national patent offices of the EEC member states.

A major step towards the conclusion of an agreement on those issues was made with the adoption of the Common Political Approach Concerning the Community Patent[14] by the Council of the European Union at its meeting on 20–21 March 2003. On each issue the Common Approach lays down the compromise solution achieved.

- The jurisdictional system would consist of a Court of Justice having exclusive jurisdiction in actions of invalidity or infringement of the Community patent rights. The Court should be established by 2010.
- The language regime provided for in the EPC would apply to the Community Patent, that is to say that applications for the Community Patent should be filed in one of the three official languages of the EPO. In this respect, of special relevance is the Agreement on the Application of Article 65 of the EPC (London Agreement) of 17 October 2000, aiming at reducing the translation cost of the European patent.
- As regards the cost for the maintenance of the Community Patent, it should not exceed that for an average European patent.

- The Community Patent system would be administered by the EPO in direct cooperation with the national patent offices of the member states. The relationship between the EPO and national patent offices would be regulated by partnership agreements which would set comparable standards of performance. Applications for Community Patents should be filed with national patent offices or directly with the EPO.
- Renewal fees would be shared between the EPO and national patent offices: 50 per cent would be kept by the EPO to cover its costs; the remaining 50 per cent would be distributed among national patent offices of the member states according to a distribution scheme established by the Council.

The Commission's Proposals for Council Decisions on the Community Patent Jurisdiction (2003)

The Common Political Approach was followed a few months later, on 23 December 2003, by two Commission proposals, one for a Council Decision Establishing the Community Patent Court and Concerning Appeals Before the Court of First Instance,[15] the other for a Council Decision Conferring Jurisdiction on the Court of Justice in Disputes Relating to the Community Patent,[16] which further developed the issue of the Community Patent Jurisdiction, outlined in the Common Political Approach. This issue is one of the most sensitive in the framework of building the Community Patent system, because of its wide territorial scope and direct impact on the national jurisdictional systems of the member states. The proposals presented by the Commission set, on the one hand, the legal basis for the establishment of the jurisdiction and, on the other, for its conferral by the Council.

Articles 225a and 245 of the EEC Treaty, as amended by the Treaty of Nice (2003), are the legal basis for the jurisdiction of the Community Patent Court, while the conferral of the jurisdiction by the Council is based on Article 229a of the said Treaty. The Commission document proposes the setting up of the Community Patent Court under Article 225a, consisting of seven judges exercising first instance jurisdiction in disputes involving the Community Patent. Within the Court, a Patent Appeal Chamber is established under Article 225(2) to hear appeals against decisions of the Community Patent Court.

2.3 Recent Developments

Since 2003, in spite of the Commission's efforts, few substantial developments have occurred in the project for creating a Community Patent system. The main result of these efforts was the preparation of a revised text[17] of the 2000 Commission proposal for a Council Regulation on the Community Patent. This text includes, on the one hand, the solutions proposed in the Common Political Approach and, on the other, those in the Commission's proposal for a Council decision on the Community patent jurisdiction.[18] However, no agreement has been reached so far on the revised text. One issue, in particular, seems to create serious obstacles to the rapid conclusions of the negotiations for establishing the Community Patent: the question of the cost of novelty searching and translating the Community patent into the languages of the EU member states.

Notwithstanding the limited results achieved, the Commission continued to hold the view that 'an affordable Community Patent would offer the greatest advantages for business', and in this context launched on 16 January 2006 a consultation[19] open to all interested persons and entities, seeking views on the system of protection of inventions in Europe and asking to indicate what changes would be required to 'improve innovation and competitiveness, growth and employment' in the Community member states.

The consultation closed on 12 April 2006. A considerable number of replies (2515 in total) were received. On 12 July 2006, the Commission convened a Public Hearing in Brussels, during which the responses received to the Questionnaire were discussed.

On the question of the 'Basic principles of the patent system', the majority pointed out that the system should be improved and that a coherent European industrial property policy should be developed. The EPO should remain as the centre of the system, supported by national patent offices.

On the question of 'Harmonization and mutual recognition', the importance of substantive patent law harmonization already achieved by the EPC and the Agreement on the Related Aspects of Intellectual Property Rights (TRIPS) (1994), was unanimously stressed, while the idea of 'mutual recognition' was generally rejected, due to the still 'unequal value' of national patents.

On the question of a 'Community Patent', the majority favoured the establishing of a Community Patent which should improve the present situation and should be a 'unitary, high quality patent'. The features of the 2003 Common Political Approach were generally rejected due to the proposed language regime and the jurisdictional system.

On the question of 'Jurisdiction', the majority supported the Community's participation in the European Patent Litigation Agreement (EPLA) in view of its 'practical and programmatic' approach and its clear procedural rules and low cost.

In his speech in Berlin on 29 March 2007, the European Commissioner for Internal Market and Services, Charlie McCreevy, speaking at a conference on 'A Europe of Innovation – Fit for the Future?', referring to the results of the 2006 consultation, stated:

> There was strong support for a cost-effective Community patent including sound litigation arrangements, while at the same time improving the current patent system in Europe. In October 2006, the European Parliament supported this line and urged us to explore all possible ways of improving the patent granting and litigation systems in the European Union. This calls for a combined effort by member states and the Community institutions.

In response to this invitation and in line with the opinion expressed by the majority of the replies received to the questionnaire, the Commission adopted on 29 March 2007 a Communication to the European Parliament and the Council entitled, *Enhancing the Patent System in Europe*.

After having noted that the present European patent system is more expensive than that of the United States and Japan, and that the existing litigation system is 'unnecessarily costly' and is the cause of 'legal uncertainty', the Communication focuses on two main issues: the Community Patent and the Jurisdiction system, in an effort to propose compromise solutions.

On the Community patent issue, the Commission's Communication underlines that the challenges Europe is confronted with in the innovation field need a patent approach that should be both 'affordable' and 'legally secure'. This means that the criticisms expressed by the majority of stakeholders on the 2003 Council's Common Political Approach, mainly because the proposed jurisdiction system and language regime were considered 'inadequate and unsatisfactory', should be seriously taken into account.

As to the translation costs, the Commission is of the opinion that it would be possible to reduce such costs, in particular to the benefit of SMEs, for example with 'fee reductions for SME or schemes allowing flexibility in the translation requirements'.

Concerning the 'Jurisdictional System', the Communication notes that opinions were expressed, on the one hand, in favour of adopting the principles of the European Patent Litigation Agreement (EPLA) in the context of the EPC and, on the other hand, supporting the establishment of a specific Community jurisdiction for European and Community patents, based on the EC Treaty jurisdictional arrangements. On this delicate issue, the Commission's Communication proposes a possible compromise solution inspired by the EPLA model and integrated by the principles of the Community jurisdiction.

In its conclusion, the Communication states that its purpose is 'to revitalise the debate on the patent system in Europe, in a way to encourage member states to work towards consensus and real progress on this issue'.

It is in this perspective that a working document entitled 'Towards an Enhanced Patent Litigation System and a Community Patent- How to Take Discussions Further' was issued by the Council on July 12, 2007, as a basis for the continuation of the negotiations on the establishment of a Community patent.[20]

3. STAKES AND OBSTACLES

Whereas there are many reasons to explain the attempts at harmonizing the patent laws in Europe, difficulties continue to characterize the history of these harmonization efforts. They may be presented as the result of technical obstacles, or of tensions stemming from contradictory interests, but also of inadequacies between innovation policies and the whole European economic policy.

3.1 Costs and Technical Difficulties

The claimed objective of the creation of a European patent was to reduce the cost induced by the filing of a patent in every chosen country. The centralized examination procedure at the European level would suppress the multiple examination procedures at the national level. The cost borne by the applicant is thus reduced. The filing and examination costs are paid at the EPO and only the grant and renewal fees are to be paid to the national offices. The examination criteria are centralized, which warrants the quality of the European patent. The latter can be demonstrated by the growing number of patent applications filed under the EPC. The number of patent applications filed with the EPO grew by 10 per cent a year on average over the period 1995–2000 and, after a downturn at the beginning of the 2000s, the growth picked up again (OECD 2006). According to the EPO, the number of total filings rose from 181 000 in 2004 to roughly 208 000 in 2006. This number of applications however seems quite low, compared with the number of applications filed with the United States Patent and Trademark Office (USPTO) and the Japan Patent Office (JPO), each of which registered more than 400 000 applications in 2006.[21] The reasons for the difference are to be found in the market attractiveness, but also in differing standards and proceedings (for example in the United States, patents are more easily granted, notably in the field of biotechnology, software and business methods; in Japan, one invention can receive several patents). Some of the reasons are also related to the shortcomings of the European patent system.

In fact, the rules governing the grant of the European patent present a number of problems. The first lies in the importance of the translation costs.

European patents filed at the EPO have to be written in one of the three official languages of the European Union (German, English and French), but if the patent is accepted, the claims have to be translated into the other official languages. To validate the patent in the selected countries, the patent must be completely translated into the language of each selected country. As a consequence, according to the European Commission, the cost of an average European patent (six countries, 18 pages, 10-year term) reaches €32 000, including €7000 of translation and related costs, which represents 22 per cent of the total cost (EPO 2006a). Of course, the higher the number of selected states, the more the translation costs would increase. This means for the applicant that the cost of a European patent is much higher than an American or Japanese one. According to the consultancy company R. Berger,[22] a patentee from an EPC member state will pay an average of €24 100 to have a Euro-direct patent granted, while a US company will pay €10 250 to have a patent granted by the USPTO, and a Japanese company will pay €5460 to have a patent granted by JPO (EPO 2006a, p. 139).

The usefulness of this translation is subject to criticism by different sources, including the EPO itself (EPO 2006b), because while its rationale is to promote the disclosure of scientific and technical information, the waiting period for the translation is very long (on average three or four years after filing). The benefits of the disclosure are thus reduced, notably in sectors where technical developments are rapid. Moreover, because of the translation costs, patents are more easily filed by big enterprises, richer in financial resources, and conversely hinder the innovation capacity of SMEs, which account for 99.8 per cent of the total number of European enterprises. Finally, the importance of translation costs induces the applicants to reduce the number of countries selected for a European patent. According to the EPO (2006a), European patents are generally valid in (only) six countries; most frequently Germany, the UK, France, Italy, Spain and Switzerland.

Another important problem in connection with the granting of European patents lies in the need to defend the patent in each country in case of counterfeiting, which increases the cost of protection under the European patent system, as well as the uncertainty stemming from decisions that may differ according to the rules in force in the different states (also due to protectionist attitudes of any state tempted to favour national enterprises).

3.2 Divergent Interests

The Agreement on the Application of Article 65 of the EPC of 17 October 2000 (the London Agreement) aiming at reducing the translation cost of the

European patent, the draft of the European Patent Litigation Agreement (EPLA) and the Community Patent, should reduce to a large extent the difficulties existing in the framework of the European patent system. The London Agreement has been ratified by France, and the relevant instrument was deposited on 29 January 2008, which means that the Agreement entered into force on 1 May 2008.

According to Lévêque and Menière (2006, p. 19), a major point to put forward in the difficult harmonization process of industrial property rights in Europe is the role played by groups opposed to the reforms:

> The most concerned interest group is the applicants, mainly composed of enterprises, for which the reforms aiming at the reduction of translation costs are positive. Other interest groups are negatively hit by the reforms. This is the case of the national patent offices and of consultants in intellectual property in charge of the designation of the European patent in the different countries. Some national applicants may be added to this category, benefiting from a protectionist judicial system. (2006, p. 19)

According to Lévêque and Menière, the enterprises' group, fragmented, large and heterogeneous, does not actively take part in the reforms, whereas the opposed groups, smaller and more homogenous, do influence the reforms more effectively. For the Community Patent, the preservation of the national offices' revenue is an answer to the influence of this interest group, while limiting its reach.

The coexistence of national, European and Community patents may create overlap in the procedures to seek patent protection in Europe and thus increase the costs borne by applicants (due to the necessity to compare the cost and profit of each system). It may also give rise to a competition between national offices and the EPO, to preserve their interests and revenues, to the detriment of innovation incentives. The complementarity between the EPO and national offices, foreseen in the Community Patent, could reduce this risk.

Finally, another argument advanced by Lévêque and Ménière (2006) refers to the creation of a specialized Court which may have an impact on the level of patent protection. The authors take the example of the United States patent system, where the creation of the Court of Appeal of the Federal Circuit has led to judgments favourable to patent holders, and thus to a stronger appropriation of invention as a consequence to widening patentability conditions (see also Gallini 2002).

3.3 European Policy

A third argument may be added to explain the current difficulties incurred by the harmonization of patent laws in Europe: that is the contrast existing between on the one hand, a restrictive monetary policy and the constraints of the budgetary policy (i.e. the Maastricht criteria and the Stability and Growth Pact) and the important element of irreducible expenses (e.g. expenses due to people ageing), and on the other hand, the ambitious aims in terms of innovation (von Tunzelmann 2004). We can recall that in 2000, the Lisbon European Council placed R&D at the centre of innovation-based policies to meet the goal of Europe becoming the 'most competitive and dynamic knowledge-based economy in the world'. To reach that aim, the Barcelona European Council set a target for European R&D intensity of 3 per cent by 2010, with two-thirds of the R&D to be contributed by the business sector, which at mid-term of the deadline seemed highly unrealistic,[23] notably because of the previously mentioned macro-economic constraints.

Moreover, it is now well known that the increase in R&D expenditures does not systematically mean better results in terms of innovation. It is the whole innovation system (its strength, clarity and organization of institutions) which has to be involved to obtain better results. However, in this beginning of the 21st century, the debates dealing with the construction of an innovation system at the European level are not at the forefront of the public agenda, as it was the case at the end of the 1990s, because of the successive enlargements of the EU, of the ways that European decisions are taken (unanimity), but also of the central place that monetary and budgetary policies have taken in the framework of the European political approach. Scientific and technological policies, which focus more on the creation of networks (interactions with local environment, users' needs) than on voluntary policies, are an illustration of this insufficient attention (in terms of budget support) given to innovation. Weak results in European scientific production and in industrial performance are the result of this contradiction (Dosi et al. 2006).

We can consider that the slowness of the harmonization of European intellectual property rules also results from the contradiction and the fragmentation of European policies. For example, in a recent book, Guellec and Van Pottelsberghe de la Potterie (2007, p. 3) argue that the reforms of the European patent system should not be an issue dealt with solely by legal scholars and that the economic dimension should be reinforced: 'The patent system will not change before legal scholars, who shape it, have acknowledged the economic dimension of their activity, and before economists have adapted their reasoning to the specific institutional and legal context of patents.' The legal approach focuses in fact on 'issues of fairness and balance of rights, of consistency of patent law with other bodies of laws' (ibid.), but does not take enough account of the 'utilitarian role of patents', which refers to their benefits for society (encouragement and diffusion of innovation) and to their costs (monopoly, restriction on use of inventions). Also, the difficult setting up of a European innovation system, in which the legal framework (including IPRs) plays a central role, can also be considered, and crucially so, as the result of the weak integration of economic and technological policies and of the insufficiently voluntary character of the European scientific and technological policy.

4. ELEMENTS OF A CONCLUSION

Despite the numerous and pertinent reasons that justify the harmonization of European intellectual property rules and the creation of an integrated patent system in Europe, various difficulties hinder this process. The latter are related to technical obstacles, to interest conflicts that in the last analysis show up the national character of this property title, conceived since its origin to promote national interests. In a context of liberalization of markets, 'neomercantilist' practices, aimed to protect and support national industries, did not disappear (Uzunidis and Laperche 2004). The difficult emergence of an integrated patent system in Europe supports this argument: making concessions in this field is considered as abandoning national sovereignty. Moreover, the orientation of European policy, built on macroeconomic equilibrium, induces the rationalization (if not the reduction) of expenditures. This leads to a focus on short-term results, seemingly a priority in the orientation of scientific and technological policies: through networks, the scientific production will quickly be transformed in marketable artefacts. But it is also the case for constructing an integrated patent system. The efficiency of patents relies on a fair balance between protection and diffusion of scientific and technological information. If the relations of power are favourable to the enlargement of patentability and to the reinforcement of the protection granted to applicants, the quality of the European intellectual property system will be reduced. A perverse effect for the promotion of innovation could ensue from this scenario, as has been stressed for the case of the United States (Gallini 2002). The harmonization of intellectual property laws and the creation of an integrated patent system in Europe are intrinsically linked to the European scientific and technological policies. To achieve positive results in this field, the latter should become a European political priority.

NOTES

1. Quoted by USPTO, http://www.uspto.gov. Abraham Lincoln is the only US President who was granted a patent (Patent No. 6469 for 'a device for buoying Vessels over Shoals').

2. The PCT is an international treaty conceived to simplify the procedure for the grant of patents when protection is sought in several countries at the same time.
3. The EPC is a regional treaty setting up a European patent system, providing for substantive rules for the grant of a European industrial property title – the European Patent – valid in all members States of the Convention.
4. Austria, Belgium, Bulgaria, Croatia, Cyprus, Czech Republic, Denmark, Estonia, Finland, France, Germany, Greece, Hungary, Iceland, Ireland, Italy, Liechtenstein, Latvia, Lithuania, Luxembourg, Malta, Monaco, the Netherlands, Norway, Poland, Portugal, Romania, Slovak Republic, Slovenia, Spain, Sweden, Switzerland, Turkey, the UK.
5. The CPC was signed in Luxembourg on 15 December 1975 by the following states: Belgium, Denmark, France, Germany (Federal Republic of), Ireland, Italy, Luxembourg, the Netherlands, the UK; it was amended on 15 December 1989 by the Agreement Relating to the Community Patent.
6. The 1989 Agreement amending the CPC included a Protocol on the settlement of litigations concerning the infringement and validity of the Community patent.
7. Green Paper on the Community Patent and the European Patent System (COM (97) 314 final, 24 June 1997).
8. The Economic and Social Council (OJ C 129, 27.4.1998, p. 8) and the European Parliament (OJ C 379, 7.12.1998, p. 268).
9. Communication on the follow-up to the Green Paper on the Community Patent and the Patent System in Europe (COM (99) 42 final, 5 February 1999).
10. Proposal for a Council Regulation on the Community Patent (COM (2000) 412 final, 1 August 2000).
11. Council Regulation No 40/90EEC of 20 December 1993 on the Community Trade Mark, OJ L 11, 14.1.1994, p. 1.
12. Amended Proposal for a Council Regulation on Community Design of 21 June 1999 (COM (1999) 310 final).
13. See, in particular: Opinion of the Economic and Social Council on the 'Proposal for a Council Regulation on the Community Patent' (CES/2001/411); Commission Staff Working Paper: 'A Community policy for the realization of the Community Patent in the context of a revision of the European Patent Convention' (SEC/2001/744 final); Commission Working document on the planned Community Patent jurisdiction (COM(2002) 480 final). See also the following revised texts of the 'Proposal for a Council Regulation on the Community Patent': the Presidency – document 8539/03; the Presidency – document 10404/03 (PI 53); the Secretariat of the Council – document 10728/03; the Secretariat of the Council – document 14233/03; the Secretariat of the Council – document 15086/03.
14. Common Political Approach Concerning the Community Patent (EU Council document 7159/03, 7 March 2003).
15. Proposal for a Council Decision Establishing the Community Patent Court and Concerning Appeals Before the Court of First Instance (COM (2003) 228 final, 23 December 2003).
16. Proposal for a Council Decision Conferring Jurisdiction on the Court of Justice in Disputes relating to the Community Patent (COM (2003) 227 final, 23 December 2003).
17. Proposal for a Council Regulation on the Community Patent (Revised Text) (Presidency of the EU Council, 7119/04, 8 March 2004).
18. See above Notes 14, 15 and 16, respectively.
19. EU Commission, *Questionnaire on the Patent System in Europe*, 9 January 2006.
20. EU Council document 11622/07, 12 July 2007.
21. httpp://www.epo.org/focus/patent-system/patents-around-the-world.html, last updated: November 2007.
22. In 2004, the EPO decided to update and complement former cost estimates. For that purpose, it commissioned a survey of applicants and attorneys which was carried out by a consultancy company: Roland Berger Market Research, see EPO (2006a).

23. The study conducted by Arundel and Hollanders for the European Union – based on the 13 countries of the European Union that account for 95.4 per cent of total business expenditures for research and development (BERD) among the EU25 countries in 2002 and 93 per cent of GDP – showed that 'the EU could only achieve a BERD intensity of 2 per cent by 2015 *if the R&D intensity of all sectors in all countries grew at the highest growth rate observed in each sector* . . . We conclude that the 2 per cent BERD intensity goal is unrealistic and unachievable by 2015. It would require massive and economically painful changes in the structural distribution of sectors within Europe' (Arundel and Hollanders 2005, p. 31).

BIBLIOGRAPHY

Aghion, P. and P. Howitt (1998), *Endogenous Growth Theory*, Cambridge, MA: MIT Press.

Amable, B., R. Barré and R. Boyer (1997), *Les systèmes d'innovation 'à l'ère de la globalisation*, Paris: Economica.

Andreff, W. (2003), *Les Multinationales globales*, Paris: La Découverte.

Antonelli, C. (2005), 'Models of knowledge and systems of governance', *Journal of Institutional Economics*, **11**, 51–73.

Arrow, K.J. (1962), 'Economic welfare and the allocation of resources for invention', in Nelson R.R. (ed.), *The Rate and Direction of Inventive Activity: Economic and Social Factors*, Princeton, NJ: Princeton University Press for NBER, pp. 609–25.

Arundel, A. and H. Hollanders (2005), *Policy, Indicators and Targets: Measuring the Impacts of Innovation Policies*, European trend chart on innovation, European Commission, accessed November 2006, at www.cordis.lu.

Bairoch P. (1993), *Mythes et paradoxes de l'histoire économique*, Paris: La Découverte.

Beltran A., S Chauveau and G. Galvez-Behar (2001), *Des brevets et des marques. Une histoire de la propriété industrielle*, Paris: Fayard.

Branscomb, L. and J. Keller (eds) (1998), *Investing in Innovation*, Cambridge, MA: MIT Press.

Cantwell, J. and S. Iammarino (2003), *Multinational Corporations and European Regional Systems of Innovation*, London: Routledge.

Caracostas, P. and L. Soete (1997), 'The building of cross-border institutions in Europe: towards a European system of innovation?', in C. Edquist (ed.), *Systems of Innovation: Technology, Institutions and Organization*, London: Pinter, pp. 395–419.

Castells, M. (2001), *La Société en réseaux, L'ère de l'information*, Paris: Fayard.

Chesnais, F. (1994), *La Mondialisation du capital*, Paris: Syros.

Chesnais, F. (ed.) (1996), *La Mondialisation financière*, Paris: Syros.

Dosi G., P. Llerena and M. Sylos Labini (2006), 'The relationships between science, technologies and their industrial exploitation: an illustration through the myths and realities of the so-called "European Paradox" ', *Research Policy*, **35**, 1450–64.

Edquist, C. (ed.) (1997), *Systems of Innovation: Technology Institutions and Organisations*, London: Pinter.

EPO (European Patent Office) (2006a), 'The cost of patenting a sample patent – new estimates', accessed at www.epo.org.

EPO (European Patent Office) (2006b), 'The London Agreement: European patents and the costs of translations' accessed November 2007 at www.epo.org/about-us/publications/general–information/london-agreement.html.

Foray, D. (2004), *The Economics of Knowledge*, Cambridge, MA: MIT Press.
Freeman, C. (1987), *Technology Policy and Economic Performance*, London: Pinter.
Gallini, N.T. (2002), 'The economics of patents: lessons from recent US patent reform', *Journal of Economic Perspectives*, **16** (2), 131–54.
Guellec, D. and B. Van Pottelsberghe de la Potterie (2007), *The Economics of The European Patent System. IP Policy for Innovation and Competition*, Oxford: Oxford University Press.
Hagedoorn, J. (2002), 'Inter-firm R&D partnerships: an overview of major trends and patterns since 1960', *Research Policy*, **31** (4), 477–92.
Hilaire Perez, L. (2000), *L'invention technique au siècle des Lumières*, Paris: Albin Michel.
Ilardi, A. (1999), *Manuale dei Trattati di Proprietà Intellettuale*, Bologne: Zanichelli.
Ilardi, A. (2004), *International Encyclopaedia of Intellectual Property Treaties*, Oxford: Oxford University Press.
Ilardi, A. (2005), *Propriété intellectuelle. Principes et dimension internationale*, Paris: L'Harmattan.
Laperche, B. (1998), *La firme et l'information. Innover pour conquéir, L'Esprit Economique*, Paris: L'Harmattan.
Laperche, B. (2001), *Propriété industrielle et innovation*, Paris: L'Harmattan.
Laperche, B. (2004), 'La propriété industrielle: moteur ou frein à l'innovation', in D. Uzunidis (ed.), *L'innovation et l'économie contemporaine*, Brussels: De Boeck, pp. 63–84.
Laperche, B., J.K. Galbraith and D. Uzunidis (2006), *Innovation, Evolution and Economic Change. New Ideas in the Tradition of Galbraith*, Cheltenham, UK and Northampton, MA, USA: Edward Elgar.
Laperche, B. (2007), 'Knowledge capital and innovation in multinational corporations', *International Journal of Technology and Globalisation*, **3** (1), 24–41.
Levêque, F. and Y. Menière (2003), *Economie de la Propriété industrielle*, Paris: Repères, La Découverte.
Levêque, F. and Y. Menière (2006), 'Le réforme du système de brevet européen: pourquoi et comment?', *Reflets et perspectives*, **XLV** (4), 11–22.
Lundvall B.Å. (ed.) (1992), *National Systems of Innovation: Towards a Theory of Innovation and Interactive learning*, London: Pinter.
Maskus, K.E. and J.H. Reichman (2004), 'The globalization of private knowledge goods and the privatization of global public good', *Journal of International Economic Law*, **7** (2), 279–320.
Mgbeoji, I. (2005), *Global Biopiracy, Patents, Plants and Indigenous Knowledge*, Vancouver, BC: UBC Press.
Nelson, R.R. (1959), 'The simple economics of basic scientific research', *Journal of Political Economy*, **67**, 297–306.
OECD (2002), *Frascati Manual*, 6th edn, Paris: OECD.
OECD (2004), *Patents and Innovation: Trends and Policy Challenges*, Paris: OECD.
OECD (2006), *Compendium of Patent Statistics*, accessed October 2006 at www.oecd.org.
Ohmae, K. (1985), *Triad Power*, New York: Free Press.
Porter, M.E. (1998), *The Competitive Advantage of Nations*, London: Palgrave Macmillan.
Rosenberg, N. (1982), *Inside the Black Box: Technology and Economics*, Cambridge: Cambridge University Press.

Schmiemann, M. (1998), 'Prospects for a single, unitary European Community Patent', *World Patent Information*, **20**, 181–3.
Schumpeter, J.A. (1976), *Capitalism, Socialism and Democracy*, Originally published 1942, New York: Harper Perennial.
Schumpeter, J.A. (2006), *The Theory of Economic Development*, Originally published 1911, New Brunswick: Transaction Publishers.
Scotchmer, S. (2004), *Innovation and Incentives*, Cambridge, MA: MIT Press.
Solow, R. (1957), 'Technical change and the aggregate production function', *Review of Economics and Statistics*, **39**, 313–20.
Tidd, J., J. Bessant and K. Pavitt (2005), *Managing Innovation. Integrating Technological, Market and Organizational Change*, Chichester: J. Wiley and Sons.
UNCTAD (2006), *World Investment Report 2006: FDI from Developing and Transition Economies: Implications for Development*, accessed December 2006 at www.unctad.org.
Uzunidis, D. (2003), 'Les facteurs actuels qui font de la Science une force productive au service du capital', *Innovations, Cahiers d'économie de l'innovation*, **17** (2), 51–78.
Uzunidis D. and B. Laperche (2004), 'Power of the firm and new mercantilism: an analysis based on Joan Robinson's thought', in M. Forstater and L.R. Wray (eds), *Contemporary Post Keynesian Analysis*, Cheltenham, UK and Northampton, MA, USA: Edward Elgar, pp. 333–48.
Uzunidis, D., S. Boutillier and B. Laperche (1997), *Le Travail bradé*, Paris: L'Harmattan.
von Tunzelmann, N. (1995), *Technology and Industrial Progress*, Aldershot, UK and Brookfield, US: Edward Elgar.
von Tunzelmann N. (2004), 'Integrating economic policy and technology policy in the EU', *Revue d'économie industrielle*, **105**, 85–104.
von Tunzelmann N. and S. Nassehi (2004), 'Technology policy, European Union enlargement, and economic, social and political sustainability', *Science and Public Policy*, **31** (6), pp. 475–83.
WTO (2006), *International Trade Statistics 2006*, accessed December 2006 at www.wto.org.

PART II

Military-based innovation networks

5. The relationship between military and commercial technologies: an empirical and analytical perspective

Claude Serfati

1. INTRODUCTION

'War is necessary for technological progress'. While this kind of claim is quite popular and often translated into academic language, it offers very little, if any help. Certainly, ever since the dawn of humanity, war has been part of social life, and for that reason it has also been a component and a driver of technological advance. This observation is still quite insufficient, nonetheless, because the huge economic, social and environmental cost – what economists narrowly call 'opportunity costs' – have to be taken into account. Beyond that, more than such a cliché is needed if we are to adopt an institutionalist-based analysis and provide an historical account of the (changing) relations between military and commercial technologies.

This chapter addresses some issues related to military and commercial technology relations. It is structured as follows. Section 1 puts the debate in its historical context, and underlines that discourses on these relations have changed over the last six decades, going from 'spin-off' to 'spin-in', then to 'dual-technologies'. In the mid-1990s, it was generally thought that generalising dual-use technologies would bring to an end the story of the relations between military and commercial technologies. The Internet case is then presented, which in the course of a history spanning four decades, provides evidence of a more complex relation between military and commercial than the usual claim that 'the Internet was invented by the military'.

Sections 2 and 3 address how the new security agenda adopted by the European Union (EU) and the USA combines in new and unexpected way, military and civilian technologies in some significant sectors, and thus opens new technological fields to the military. Section 2 documents how the space sector has become closely connected to security technologies on the EU agenda. From a military–civilian technology relationship, European space programmes also offer an interesting insight on how security is becoming a framework, if not an umbrella word, for the development of technologies used in an integrated way for military and civilian purposes. Drawing on the US experience, the case of biotechnology addressed in section 3 is interesting as it is a set of technologies which have been mainly oriented to civilian markets since the end of World War II, and kept in a (relatively) low position by the military. The recent involvement of the Department of Defense (DoD) and particularly of the Department of Homeland Security (DHS) further indicates the enlargement of the perimeter which is now considered as relevant for military. This is also the case for the space sector in Europe. Findings from sections 2 and 3 are that through the security framework, a new era has opened at the turn of the century as far as military and commercial technology relations are concerned. In the conclusion, a synthesis based on case studies addressed in the previous parts of the chapter is offered, leading to the proposal of five criteria which are relevant for an investigation of the relations between military and commercial technologies.

2. MILITARY–CIVIL RELATIONS AND THE INTERNET CASE

2.1 'Spin-off, Spin-in, Dual-technologies': Changing Discourses

As already stated, claims such as 'War is necessary for technological progress' offers little help.[1] There is plenty of anecdotal evidence of artefacts which, once designed for warfare purposes, diffused into the civilian markets,[2] but conversely some reach quite the opposite conclusion, going so far as to claim that 'to the limited extent we can generalize about five hundred years of history, it seems fair to say that the most radical innovations come from outside of formal military structures' (Boot 2006, p. 457). When the historical and institutional background is integrated into the analysis, there is a no doubt that World War II marked a turning point as far as the relations between military and commercial technologies in capitalism are concerned.[3] The 'Manhattan Project', spanning 27 months – from the spring of 1943 to fall of 1945 – employed over 130 000 people and cost a total of $1.9 billion (around $23 billion in 2007 dollars). Instead of what happened after World War I, when the US government funding retreated, the post-World War II era witnessed an increasing role of federal, mainly defence-oriented, funding. The Manhattan Project made 'Big Science' come of age, even though it had begun during the 1930s.[4] Big science is characterised by: 'hierarchical organization and management structures and committees; strong interactions between basic science, technology and engineering; complex patterns of extra-mural industrial and government funding and influence, and the national and local prestige derived from large instruments and large-scale funding as well as from scientific achievements' (Freeman and Soete 1997, p. 376). In the post-war decades, Big Science acted as a general framework for major technological trajectories, and did not restrict its mode of operation to the aerospace–electronics–nuclear complex.

Among scholars there is an agreement, which more or less became conventional wisdom, that in the two decades following World War II technologies designed for military purposes spilled over into the civilian economy. More precisely, the positive role of the military in the development of technologies can be accounted for by the funding of unexplored fields, the creation of the market through procurement, the support given to domestic firms. Success stories include semiconductors, e.g. solid-state transistors and integrated circuits (ICs), where the military supported civilian firms through research contracts and massive demand (the first major military application of ICs was in the Minuteman II guided missile).[5] In computer software, Flamm (1988) finds that, of 45 advances between 1950 and 1980 which originated in the United States, the development of 18 was funded by the DoD. Strong transfers also happened in the aircraft sector – with the Boeing B-47 metamorphosing into the Boeing 747 – and jet engine propulsion (Germany had first developed the jet engine and its experimental aircraft, the Heinkel He-178, flew in August 1939). The space industry also closely connects military and civilian spheres, and cases of spin-off from the military are numerous. They include rockets (again it was the Germans who first used propulsion guided missiles, though other countries possessed the technological capabilities), civilian space launchers (developed after their military counterparts) and ballistic missiles. Obviously, nuclear power for civilian purposes was directly derived from military needs-oriented research. Not only technologies, but also engineering sciences developed thanks to military support: nuclear physics, cryptography and cybernetics are only a few examples.

Conversely, failure stories are numerous in other or the similar sectors, including aerospace, with the attempts to derive a supersonic commercial aeroplane from military supersonic applications by Boeing (SST) and a Franco-British cooperation (Concorde). In electronics, the US military R&D in semiconductors has generally not produced basic technological innovations (de Grasse 1984). The US Very High Speed Integrated Circuit (VHSIC) project and the Strategic Computing Initiative, two large projects largely funded by the military, both failed (Stowsky 2004). The VHSIC aimed to develop advanced computer and data-processing architecture, new approaches to computer-aided design, and research into material and physical process needed to achieve submicron geometries. It was because of the increasing dependence of defense systems and sub-systems on foreign electronic components that in 1987 the Congress authorized SEMATECH, a consortium formed between the federal government and industry. The federal government, mainly through the DoD, proposed to contribute up to 50 per cent of the cost over a 5–10 year period for development of new technologies. The Strategic Computing Initiative, launched by the Defense Advanced Research Projects Agency (DARPA) in 1983, was a plan to develop machine-intelligence technology and recent technological advances and opportunities in microelectronics, computer science and artificial intelligence (AI). A major reason for failure was because the project funded development only of military-specific applications of the technologies that the commercial firms would not otherwise have developed themselves. These projects replicated the project characteristics that had failed to foster commercial advances (or worse, distracted companies from pursuing them). Another sector severely hurt by military priorities was the development of numerical control technology for machine tools. Ergas (1987) has documented how the development in the 1950s of numerically-controlled machine tools at the Massachusetts Institute of Technology with the funds of the US Air Force led to machinery that offered a strong centralization of control and wide versatility, while ignoring cost constraints.

In France likewise, success and failures are both present. There is no doubt that intra-sectoral transfers occurred in aeronautics and space, and also from the lack of significant competitors on the world market, in the nuclear industry. In those sectors, military R&D and procurement were able to boost French companies at the world first-class level. The success of EADS (parent of Airbus company) in jet aircraft, SNECMA in aircraft engines, Ariane and the European Space Agency in space, Framatome in boilers, and COGEMA in recycling processes are undoubtedly a combined effect of a long tradition of excellence in the related scientific fields, and a consequence of strong military R&D and procurement. Costly failures are at least as numerous as successes. The 'Plan Calcul', heavily state-funded in the 1960s to provide computing capabilities needed to design and develop the nuclear bomb, did not succeed in developing a civilian market-oriented industry. Another sector hurt by military R&D priorities is that of 'equipment goods'. Aggregate sector-level data somewhat mask the fact that some 'equipment goods' sub-sectors which are not defence- or aerospace-oriented and which either are located at the core of technological and industrial linkages, e.g. mechanical engineering[6] and electrical, or are strongly consumer-oriented (domestic appliances), or are both (computers), have been steadily losing ground over the two last decades (Serfati, forthcoming).

As it appears that over the three post-World War II decades, the record of successful spin-offs from the military to the commercial sector could be matched by that of failures, the debate changed in the early 1980s. A number of factors set the tune for another discourse, one around 'spin-in' or 'spin-on', meaning that the flow of technology had moved the other way round, going from the commercial to the military sector. To begin with, spin-off claims were challenged on the evidence of the loss of international competitiveness in manufacturing industries by countries endowed with large military budgets (and strong military R&D) to the benefit of countries with smaller military budgets. The USA, the UK and France were in the first group, as Germany, Japan, Sweden were in the second. The loss of market share by the USA, UK and France in fast-growing and core industries, in particular electronics equipment goods was attributed to the priority placed on military programmes in their 'mission-oriented' technology policies (Ergas 1987). It became clear that those technologies designed for commercial customers had upgraded to the point that they could be integrated into weapon systems. In the late 1980s, their lower cost, improved reliability, coupled with budgetary constraints on defence expenditures, made them all the more appealing.

This intermediate phase, roughly spanning the 1980s, was followed by discourses and policy recommendations on promoting 'dual-use technologies' (DUT). Cowan and Foray (1995) proposed to distinguish between technological spillovers[7] and dual-use technologies. Technological spillovers occur when 'particular research is done exclusively in one domain and adopted more or less without change in other', while 'dual-use technologies describe technologies that are developed and used by the military or space sectors on the one hand and by the civilian sector on the other'. They showed that various types of transfers exist: direct transfers of specific components, indirect transfers (incorporation of results into improved products and process), informational effects (from basic research), organisational effects (maintaining a critical mass of scientist, diffusing principles of management, etc.) (ibid., p. 854). A different definition, and somewhat more challenging analysis, is proposed by Molas-Gallart (1997, p. 370): a technology is said to be dual-use when it has *current* or *potential* military and commercial applications (emphasis added).

The dual-use debate was useful because it departed from overgeneralisations on spillovers and spin-offs – with sometimes a flavour of enthusiasm for the role of the military – and looked to open the black box of intersectoral technological transfers. The downside is that in the debate on dual-use technologies, theoretical and policy decision-making perspectives were closely linked, making it difficult to disentangle academic analysis and normative recommendations. In the United States, under the Clinton administration, technology policy to promote dual use came out of the closet, and at the Pentagon, Secretary of Defense William Perry introduced an official policy for the increased use of civilian technology. The Technology Reinvestment Program (TRP), launched in 1992, although subsequently killed off by a new Republican Congress (1995), survived through a 'Dual Use Applications' programme. In Europe in the early 1990s, the French procurement agency (DGA), in coordination with the European Commission (see below) was keen to promote dual-use technologies, as it was convinced that, given the overarching role of the defence innovation system in its national system of innovation, such policies would leverage its influence (Serfati, in Gummett and Stein 1997).

The dual-use era was generally thought of as bringing closer together defence- and commercial-related innovation networks. More optimistically, from these analyses, some concluded that the military-industrial complex (MIC) that had thrived during the Cold War period thanks to its 'entrenchment' would be progressively absorbed into and become a part of a broader domestic technological base. The 'new economy', civilian market-driven, added further momentum to this conjecture. Still, less than a decade later, the landscape appears rather different. In the USA, the new geopolitical and economic context resulted, not in an attrition but in a regeneration (or recreation) of the MIC. In France, the 'Meso-system of armaments', surprisingly resisted drastic 1990s budgetary cuts and the steady loss of France's geopolitical influence in international relations. French defence contractors continue to hold a core place in their national system of innovation and technology policy, as their role in technology matters overwhelms their overall place in the manufacturing industry as a whole (Serfati, forthcoming, Table 5.1).

2.2 The Internet and the Military

The role of the DoD in the creation and the development of the Internet is well documented (see, for example, Abbate 1999). The story began with a small project after the tensions between the USA and USSR reached a climax in 1962 (the Cuban missile crisis), when a nuclear confrontation seemed imminent. In this historical context, the military's initial goal was more prosaic. Administrators at DARPA wanted to save money on duplicative computer purchases. One of the founders of the Internet, Paul Baran, a researcher at RAND – a research organisation created after World War II to help maintain the technological edge gained by the US military during the war – offered a solution: design a more robust communications network using 'redundancy' and 'digital' technology. In 1969, the Advanced Research Projects Agency (created in 1958 as ARPA, then DARPA in 1972, then ARPA in 1993 and again DARPA in 1996) funded an experimental packet-switching network. In 1970 this network, called the

Table 5.1 Defence-related indicators compared to macro-economic, manufacturing, export, and technological national indicators

Indicators	Share (in %) or absolute number
1) Macro economic indicators	
Military expenditures/total public budget	11.5%
Military expenditures/ GDP	1.8%
Aggregate arms export surplus/aggregate equipment good surplus (1990–2006)	63%
2) Manufacturing industry indicators	
Defence consolidated turnover/manufacturing industries turnover	18
Defence industry employees/manufacturing industries value added	6.8%
Number of defence-aeronautics-nuclear companies in the top 10 companies of the equipment good sector	7%
3) Related technological innovation indicators	
Public defence R&D/total business R&D (2004)	24%
Total R&D spending by defence-aeronautics-nuclear companies/total R&D spending by the top 20 R&D French companies spending (2005)	24%
Number of R&D defence-aeronautics-nuclear companies/ top 20 patenting companies in France	20%

Source: Author, from various sources

'ARPANET', linked five 'nodes'. Then, in a couple of years ARPANET was converted from an experimental network into an operational network, and the responsibility for administering the network was given to the Defense Communications Agency (DCA, now DISA: Defense Information Systems Agency). In 1983, major changes took place, with the adoption of the TCP/IP protocol by DARPA and the split between ARPANET and MilNet, dedicated to military purposes. In 1982–83 came CSNET, established by the National Science Foundation (NSF), followed in 1985 by the creation of NFSNET, which became connected to the then-existing Internet. Because of its higher speed, NFSNET became the 'backbone' of the future Internet, replacing ARPANET which formally disappeared in 1989. In 1991, the introduction of Hyper-Text Markup Language (HTML) and Hyper-Text Transfer Protocol (HTTP) by Berners-Lee and Cailliau at CERN opened the way to the World Wide Web (WWW). The extraordinary development of commercial infrastructures during the 1990s led the NSF to dismantle NSFNET in 1995 which was replaced by today's commercial Internet.

Based on this brief overview, a number of comments are in order. First, ARPA played a leading funding role. Over the 1960 and 1970 decades, funding by ARPA overtook NSF funding by a large margin (Langlois and Mowery 1996). According to their estimates, between 1956 and 1980 the cumulative NSF funding for research in 'software and related areas', amounted to more than $250 million (1987 dollars), most of this funding going to US universities, as the DARPA R&D funding from its Information Processing Techniques Office (IPTO), which went to both universities and industry, averaged roughly $70 million annually (1987 dollars) between 1964 and 1980. Still, it can be said that ARPA acted as a 'catalyst' of research already undertaken by scientists at RAND or in universities. By a catalyst is meant that one could say that this 'idea was in the air', to paraphrase Alfred Marshall. This concerns two major technologies developed thanks to ARPA funding, the packet-switched communications and the TCP/IP. At the time when ARPANET was designing its prototype packet-switched network, other research teams, independent of the latter, worked on similar issues, in the USA (National Physical Laboratories), in the UK and in France (CYCLADES built in 1972), the latter introducing significant technical changes, but running out of funding in 1978 (Mowery and Simcoe 2002). Likewise, the TCP/IP, which has become the communication standard for the Internet, was developed by Cerf and Kahn, two DARPA-funded engineers in 1972 or 1973.[8] Again, almost at the same time, CERN (Conseil Européen pour la Recherche Nucleaire) was involved in similar research aimed at easing communications between different techniques and protocols existing because of strong competition between proprietary systems, but also because of home-made systems. CERNET, developed at CERN and beginning in 1976, offered a fast file transfer service between mainframes and minicomputers. According to one of its creators at CERN, even though architecturally it resembled ARPANET, all its protocols had been developed independently (Segal 1995). It is not underplaying the role of ARPA to observe that research on similar themes was carried out in different places, including in those not funded by the military.

Similarly, other technologies, critical in the very early stages of the development of the Internet, were not funded by ARPA. They include the computer operating system: UNIX was created at Bell Labs in the same year (1969) that ARPANET was created. The former was thus not an outcome of the latter. Also, Ethernet, a local area network (LAN) technology that transmits information between computers, was developed by Robert Metcalfe at the Xerox Palo Alto Research Center (Xerox PARC) between 1972 and 1976.

Second, the respective and interacting role of defence and civilian agencies has to be carefully looked at if we are to address relations between military and civilian technologies. In the early stages of the process (late 1960s), ARPA funds were channelled through the US Navy, and serious ARPANET research at that time and for a decade afterwards often required security clearance. Then, ARPA acted to encourage academics and research centres to bring new solutions, encouraging a kind of emulation among the academic 'centres of excellence' (MIT, UC Berkeley, Stanford, Carnegie-Mellon, the University of Utah, and UCLA) (Mowery and Simcoe 2002) and, as seen above, it was also the main funding public institution. Still, it seems that this open attitude adopted by ARPA was at some point reluctantly accepted by other parts of the military establishment which feared that a diffusion of the new technology would imperil national security interests. Because it was mindful of the risk of dwarfing research, as had been the case in other industries (semiconductors, machine tools), in July 1975 DARPA transferred management and operation of the ARPANET to the Defense Communications Agency (DCA). The DCA's main objective was to focus on military applications and prevent the network from being exposed to hacker attacks. The DCA was concerned about preventing use of the network for 'frivolous activities' (Hughes 1998, p. 136, cited by Ruttan 2005), and it had even initially planned to dismantle ARPANET once the message switching network AUTODIN II that the DCA had developed for exclusive military use, became operational. In that context it is clear that the coexistence of military users and academic researchers was somewhat uneasy under DCA authority (IEEE, n.d.). The DCA's attitude was no longer defensible, and when it appeared that the growth of the civilian network was so overwhelming, the complete separation of the two systems was carried out in April 1983. The same year, ARPANET ceded jurisdiction over deployment to the civilian user-developers on whom the Pentagon depended for technical expertise. This was the final stage which enabled civilian users not only to use, but to monitor the developments of the Web. To sum up, the separation between the military (MILNET) and the free-access (TCP/IP ARPANET) networks, with the 'entrenchment' of the former into pure military functions, marks the very limits of the notion of 'spin-off', and paved the way for an extraordinary extension of the already fast-growing civilian network on a world scale. This was fair recognition that only an open civilian environment could promote the development of these new technologies.

Indeed, even in the early stages spanning the 1970 and 1980 decades, ARPA was not the only federal agency to promote the research. Overall, the role of the NSF can hardly be underestimated, even in the early period (1960–70). The NSF was so active and keen to support the development of networks that representatives from ARPA, NSF, and computer scientists from six universities met in May 1979. The objective was to allow US universities, most of them doing no research with the Department of Defense, to connect non-ARPANET computer science departments. The establishment in 1982–83 of CSNET was probably a turning point in the history of the Internet because the stated objective was to extend the network to every scientist and engineer in the United States. The NSF was given the responsibility of building and maintaining the trunk lines or 'backbones'. This set up an explosive growth of the Internet (from a couple of hundred computers on the network in 1983 to over 1.3 million computers in the system in January 1993). This growth was made possible by the decision by the NSF to adopt TCP/IP as the standard on its national university network, leading to the eventual dominance of this protocol (Mowery and Simcoe 2002). Also, the router technology adopted in the early 1980s to address the explosion of communications, was designed under NSFNET and funded by the NSF, as the technology used in ARPANET was too rudimentary (routers were custom-made on a case-by-case basis), because the research firm (BBN) which developed the first routers for ARPANET did not believe there was a commercial market for them (Cerf 1993). The technical complexity of routing was thus increased substantially under the open system of NSFNET. Another indication of the core role of the NSF in the technological development is the fact that, according to Vinton G. Cerf, the co-creator of TCP/IP, CSNET adopted TCP/IP, 'but developed a dial-up "Phone-mail" capability for electronic mail exchange among computers that were not on ARPANET, and pioneered the use of TCP/IP over the X.25 protocol standard that emerged from commercial packet switching efforts' (Cerf 1995). Finally, as NSFNET strove to promote the development of the Internet, it was also assisting its commercial development by giving the promise of time-limited support, after which the service provider would have to be self-supporting (Cerf 1995).

Third, and by way of synthesising this discussion, based on this historical account, claims that 'the Internet was invented by the military' or even 'would not have existed without military support' has little relevance from an evolutionary perspective. Speaking of the 'invention' of the Internet is not appropriate, as this invention spans three decades, from the mid-1960s to the mid-1990s. The coming of age and the development of the Internet was the result of many scientific and technological breakthroughs, some funded by the military, others by public civilian agencies, and realised by public research (universities, research centres) and private companies. The evolution of the Internet sheds a more complex light than the 'spin-off'/'spin-in' debate. Suffice to recall that some critical computer technologies, including Ethernet, UNIX, HTML/HTTP, designed in 1989 and 1991 by Tim Beners-Lee at CERN, and of course the WWW which resulted from the latter inventions, were designed outside a direct ARPA role. At later stages, the crossing of the Internet with other technologies, e.g. communication satellites, which was so important in the growth of the Web at the end of the 1990s, or in the late 1990s the emerging of the Internet through mobile phones and 'triple play' (provision of the Internet, television and telephone on a single broadband connection), which is considered as opening a strong and long-term growth in communications, had little if anything to do with the military.

A related question on the role of the military is why the United States overtook other countries in the development of the Internet. Both matters are sometimes (and mistakenly) conflated, and the 'US model' narrowed down to the overarching role of the military in technology policy. As documented by Mowery and Simcoe (2002), beyond the role of the DoD, it is the unique features of the post-war US 'national innovation system' (institutions, policies, funding) which account for the fact that the Internet was primarily a US-based phenomenon. A key factor was the competitive advantage obtained by the size – in geopolitical and economic terms – of the United States. The US networking initiatives served as an effective deterrent to the creation of smaller regional alternatives, as the 'network effect' created by a growing network that already linked the extremely active American computer science community led international researchers to join the US effort.

3. GLOBALISATION AND THE EMERGENCE OF SECURITY TECHNOLOGIES

The transformation of the defence agenda into a security agenda by the USA and the EU, both separately and jointly through NATO, which could be perceived in the late 1990s, has accelerated since the turn of the century. Table 5.2 documents the broad range of changes which have taken place from the 1990s onwards.

Though the emergence of security issues at the core of the globalisation process has been paid great attention by political scientists and resulted in an extensive literature in the academic field, there is a remarkable lack of scholarly studies by scholars of economics of innovation and technical change. This is surprising, since the security agenda provides a doctrinal framework against which new technological trajectories are emerging, marking a new stage after the spin-off, spin-in and dual-use stages. Put otherwise, the 'security economy' is coming of age (OECD 2004). What is appealing about 'security' is that many fields and issues can be included in

Table 5.2 From a defence to a security agenda in the era of globalisation

	Defence 1945–c.1990	Security c.1990–2006
Geopolitical setting:	Two superpowers for the world order, and a Cold War (and also numerous wars)	One superpower, and an increase in chaos and world disorder (fragmentation/hierarchisation)
Objectives	***External*** *threats: military*, coming mainly from hostile states: ***Internal*** *threats:* criminality, social unrest	Intertwining of external and internal insecurity: military, environmental, natural resources, criminality, migrations, social unrest
Military capabilities	Monopoly of legitimate coercive violence (external and internal)	Monopoly of legitimate coercive violence; rise of private transnational authorities (violent groups, military private companies etc.)
Provision of the 'public good' defence	Nation-states, UN in limited conditions, NATO for the transatlantic bloc's members	States, private authorities (agencies of states, acting as 'principals', UN, NATO (global security organisation?)
Economic setting:		
Macroeconomic (public)	Anti-cyclical (Keynesian) role of defence budget	Compression of military spending because of public deficits constraints (1990s), then rise when threats are invoked (2000s); emerging of a 'security economy'
Industrial	National champions, managerialism (Galbraith's technostructure)	'Transatlantisation' of arms production, shareholder value-based governance of defence prime contractors
Technological	Diffusion from military to civilian (1950–70), then from civilian to military (1970–1990)	Integration of technologies information and security technologies
Economic theory	Defence as a 'pure' *national* public good (Samuelson, Stieglitz)	Security as a *global* public good (Kaul, Sandler, etc.)

Table 5.2 (continued)

	Defence 1945–c.1990	Security c.1990–2006
Epistemology	Of defence: Separation of disciplinary fields (economics, international relations, political sciences, sociology, etc.)	Of security: Need for interdisciplinary research (economics, international political economy, political sciences, sociology, anthropology, etc.)

Source: Author.

this framework. Actually, security-oriented technologies open up new avenues well beyond 'classical' defence-oriented sectors (aeronautics, space, and defence electronics). The private security industry is estimated at $120 billion. A growth in turnover in the order of 7–8 per cent annually is expected, easily outstripping the average rate of economic growth (OECD 2004, p. 9). In the USA, estimates are that the homeland security business will grow from approximately $40 billion in 2004, to nearly $180 billion by 2015 (US Department of Labor, 2007).

This does not just apply exclusively to the USA. The security economy is taking on a global dimension with huge economic stakes, as evidenced by the report released by the World Economic Forum (2007) (see Figure 5.1).

To the sharp rise in private spending on security has to be added public security-oriented expenditures, which are distinct from military expenditures. Again, the United States forges ahead by a large margin, with an impressive growth of the DHS budget (Figure 5.2). Funding increased drastically between FY2001 and FY2007, from $16.9 billion to at least $58.2 billion – a 246 per cent increase. The total funding appropriated by Congress between 2001 and 2007 amounts to $271.5 billion.

Biotechnology, which had been kept out of large technology programmes funded by DoD, has become a much more significant item in the DHS R&D budget. Before addressing this new and promising field for the military, we address the way Europe has taken on the security agenda.

3.1 EU's Increased Momentum in Security-related Technologies

Throughout the 1990s, the European Commission has emerged as an actor in defence issues. Even though its role was modest, this marked a

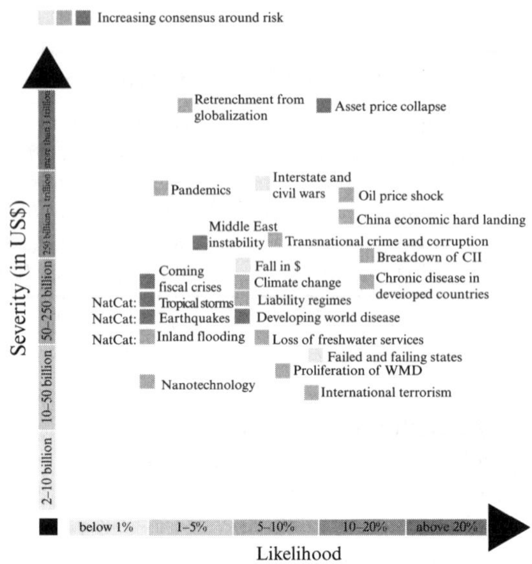

Source: *World Economic Forum Report*, 2007

Figure 5.1 *The 23 core global risks: likelihood, with severity by economic loss*

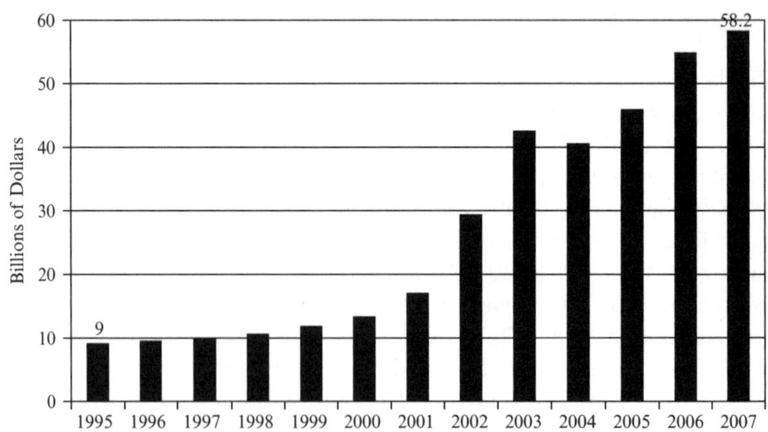

Note: Levels for FY1995 through 1997 are estimated, as the Office of Management and Budget did not collect data on these activities prior to FY1998.

Source: Budget of the U.S. Government, Fiscal Year 2007

Figure 5.2 *Trend in homeland security spending between FY1995 and FY2007*

breakthrough since the Article 296 (ex 223) of the Treaty on European Union formalised the Community's exclusion of defence from its joint responsibilities. Through dual-use technologies issues, the Commission felt it could enter the field of defence matters: it was no longer possible for the Commission to be excluded from the development of some technologies with dual-use potential (Gummett and Stein 1997).

Evidence on how the security agenda is dramatically reshuffling the technology policy landscape in Europe can be documented. The Commission accelerated its interest by bringing together space – where a long-established European collaboration exists with the European Space Agency (ESA) – and security. Its involvement developed in a number of steps, as follows. First, in July 2002, the *Strategic Aerospace Review for the 21st Century* (STAR 21 Report for short) was released by a group consisting of seven executives of the main European defence and aerospace companies, six members of the Commission (including Javier Solana, then EU High Representative for the Common Foreign and Security Policy) and two European Parliament members. The recommendations of STAR 21 cover world markets, the operating environment, governance of civil aviation, European security and defence, and space capabilities. The objective of STAR 21 was to develop a fully European-based space defence and security capability for surveillance, reconnaissance, command/control including telecommunications and positioning (STAR 21 2002, p. 39). Second, in February 2004, the Commission launched a Preparatory Action entitled *Enhancement of the European industrial potential in the field of security research 2004–2006*, with a €65 million budget, the objective being preparing the basis for a fully-fledged European security research programme starting in 2007 (Commission, 2004). Third, on 15 March 2004, the Group of Personalities (GOP) for Security Research presented its Report to European Commission President Romano Prodi. The report, titled *Research for a Secure Europe*, describes the need for increased coordination in security research, outlining 12 recommendations for the future, including a minimum of €1 billion annually for security technology development. Fourth, the GOP that gathered in April 2005 to prepare the financial plan 7 (FP7) (2007–2013) proposed to draw on the US approach, in particular in that 'the US budget illustrates perfectly well the blurring of the distinction between military- and non-military security functions and capabilities.' It continued, 'There is no reason why European security research should not be funded at a level similar to the US. Calculated as a per capita investment, the US spends more than four dollars on security-related S&T for each citizen. Using the same reasoning, this would mean that an overall EU security R&T budget of €1.8 billion for 450 million Europeans would be desirable' (Research Group, 2004, p. 27). In its conclusion, it also recommended that, in order to narrow the gap in security R&T,

and to complement national and inter-governmental efforts, the minimum threshold of a Community-funded ESRP should therefore be €1 billion per annum.

Fifth, the recommendations reached policy-makers and the Commission. Following the traditional interactive process between the European Council and the Commission, Security Research, as a part of the 7th EC RTD Framework Programme (FP7, 2007–13) received a total budget of €1.43 billion for space (of which 85% will be allocated to Global Monitoring for Environment and Security, GMES) and €1.35 for security. This represents 8.5 per cent of the €32.413 billion 'Cooperation' programme. Since the Euratom Treaty receives an additional budget of €2.751 billion, space, security and nuclear will receive 10.8 per cent of the total FP7 budget.[9] Security-oriented research (as distinct from development) is to increase by €200 million over the 2007-13 period, up from €15 million in 2004 and €25 million each in 2005 and 2006.

Details of the FP7's allocation of funding and categories of participating institutions would allow an estimate of the weight of defence and security concerns. It has been argued that under 'FP5' which ran from 1998 to 2002, 8 per cent of the total number of participants in the BRITE-EURAM (industrial and materials technologies), ESPRIT (international RTD co-operation in IT), ACTS (advanced communication technologies) and TRANSPORT programmes were military organisations (Hayes 2006).

3.2 Space and Security

Not only is space a domain where civil and commercial technologies are strongly interconnected,[10] it has also been systematically associated with the new security agenda in the language emanating from Brussels. Indeed, space is undoubtedly the most relevant field for the Commission in its quest for an increased role in security, and beyond that defence matters. European cooperation in space is, as in aeronautics, an old story, but it had mainly proceeded out of the control of the Commission and through the European Space Agency (ESA), which is not a Community-based institution. The situation changed when in 2000 the Commission and ESA set out a joint European Strategy for Space and created an EC/ESA taskforce. Links were strengthened through the two key programmes, the Galileo global positioning system and Global Monitoring for Environment and Security (GMES) launched in 1998. A Community agency, called the European GNSS Supervisory Authority was set up in 2004.[11]

From a technology policy vantage point, the Galileo programme is important as it is the largest industrial project ever organised on a European scale, and will constitute the first truly European public infrastructure, while also introducing new management practices based on public–private partnership for the first time at the European level. It has also gained a worldwide dimension, thanks to a series of cooperation agreements signed with China (30 October 2003), Israel (13 July 2004), Ukraine (June 2005), India (7 September 2005), Morocco (8 November 2005), and South Korea (12 January 2006). Several other countries, including Russia, Switzerland, Canada, Australia, Saudi Arabia and Brazil, reaffirmed their interest in continuing discussions with the European Union. The programme has already accumulated a delay of five years with regard to the initial calendar and is currently facing a number of difficulties (in 2007, initial operational capability were forecast by end-2011 and the full deployment by the end of 2012).

It became apparent that the programme was plagued with serious financial and management troubles when the Galileo consortium member companies (AENA, Alcatel-Lucent, EADS, Finmeccanica, Hispasat, Inmarsat, TeleOp and Thales) signalled in 2006 that they were backing off from their engagement in the European alternative to the US Global Positioning System (GPS), arguing they would need to bear most of the financial risks without any clear guarantees of return on their investment. The Commission reacted with a communication stating that the quickest way to make the Galileo navigation system operational was to find some €2.4 billion extra of public money to finance the construction of the infrastructure (Commission, 2007).

From the perspective of a military–civilian technology relationship, European space programmes also offer an interesting insight on how security is becoming a framework, if not an umbrella term, for the development of technologies used for military and civilian purposes. The two major European programmes, Galileo and GMES, are clearly security-oriented. This could be a compromise, quelling the fears of those opposed to a military use of European space capabilities. On this side, one can find the US Administration, along with some European countries including Britain, who have made it clear that they would oppose any kind of military use. On the other hand, France is keen that space capabilities be prepared for military tasks. The five services provided by Galileo, a system composed of a constellation of 30 satellites, highlights its military potential through the Public Regulated Service (PRS) (see Box 5.1). As the US–EU agreement was not publicly detailed, it seems that the European side accepted a slight downgrading of the signal, in order to allow, during a military conflict, the US to be able to jam all public satellites so that its enemy could not use satellite positioning.[12]

Despite the compromise arranged with the USA and recent financial

BOX 5.1 GALILEO'S FIVE SERVICES

Galileo will provide five services

- The Open Service (OS), similar to the standard civilian GPS signal;
- The Commercial Service (CS), which will provide additional information such as increased positioning accuracy;
- The Safety of Life Service (SOL) which will contain an integrity signal, essential for all applications where knowledge of (and confidence in) the integrity of the navigation system is essential;
- The Public Regulated Service (PRS), encrypted and dedicated to government-controlled applications;
- The Search and Rescue Service (SAR), able to relay distress signals transmitted at 406 Mhz by air, sea or land-bound distress beacons.

troubles, space undoubtedly has some leverage, allowing the Commission and beyond it the EU to have a greater say in security and defence issues. Once timid on addressing potential military uses by the European GPS in the early 2000s, the Commission no longer hesitates to refer to this opportunity for Galileo.[13] The case of GMES is rather similar to Galileo. It is designed for meeting missions as defined by the common foreign and security policy agenda (humanitarian and rescue tasks, and peacekeeping and peacemaking tasks of combat forces). Its military purposes become evident, as it is also expected to support national military satellites. In particular, GMES has significant intersection with the Common Operational Requirement concept (known by the French acronym BOC, Besoin Opérationnel Commun) (Pasco 2006). BOC is co-signed by six European countries (France, Germany, Greece, Italy, Spain, and Belgium) and built on national (or bi-national) domestic capabilities (Germany's SAR-Lupe, Italy–France's Cosmo Skymed-Pleiades, and France's Hélios). It will address future military needs in the field of Earth observation, and it is seen as the first step towards autonomous European capacity in strategic imagery, providing for a military satellite observation system (Assembly of Western European Union 2006). The role of GMES further exemplifies the fact that Europe is accumulating military-oriented space capabilities, raising fears that it will embark on a military race in space, as expressed in a document produced by a Committee of the European Parliament.

4. BIOTECHNOLOGY, A NEW HORIZON

Biowarfare provides further evidence of the need to contextualise relationships between technology and war. Even though since the dawn of human civilization, poisons have been used for assassination, a turning point was reached at the late nineteenth and early twentieth centuries, based on scientific discoveries and transformations of warfare under capitalism.[14] The last century has extensive evidence of massive use of bioweapons. During World War I, more than 1.3 million victims were injured or killed, 100 000 of which were killed due to chemical weapons. Even though the Geneva Protocol of 1925 prohibiting the use of chemical or biological material in war marked some progress,[15] it did not prevent further use of this kind of weaponry by leading countries. After World War I, chemical weapons were employed by the Italians in Abyssinia, by the French and Spanish in Morocco, by the Japanese in China, and possibly by the British in Afghanistan (Boot 2006, p. 451). During World War II, all the combatant countries, including Canada, France, Germany, Great Britain, Japan the Former Soviet Union (FSU) and the USA had some type of biological weapons programme (Geissler and van Courtland Moon 1999).

Despite that situation, since the end of World War II, military-funded research in biology has been relatively limited in industrialised countries endowed with strong military R&D budgets (in particular France, the UK and the USA). Main technological fields covered included nuclear, aeronautics and space, and electronics. A mix of revulsion for such weapons, the persuasion that more efficient types, including nuclear, maintained some superiority, the fear that a biological weapons programme could easily be duplicated by enemies and thus raising a threat to their own national security, are some of the reasons for the low level of military funding. This also explains why the USA, UK and FSU together supported the Biological and Toxin Weapons Convention (BWC) of 1972, the first arms control agreement to ban outright an entire class of weapons, even though the BWC does not have formal mechanisms to monitor or enforce compliance. This does not mean that biotechnology was not publicly funded, but in the USA and other countries, funding came from civilian institutions that were quite instrumental in the development of new drugs and radical discoveries.

This brief historical account is a backdrop to the sea change which is under way as far as relations between military and civilian technologies are concerned. There is no doubt either that the increased interest for bioweapons has accelerated since 9/11. This section will focus on the United States, as this country is obviously the forerunner in the technological race to new weapons. There has been a steep increase of

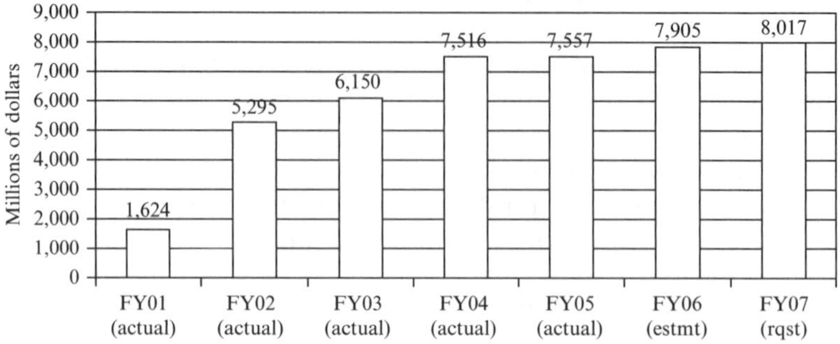

Source: Center For Arms Control And Non-Proliferation, Biological And Chemical Weapons Control Program, 2007

Figure 5.3 Total annual federal funding for bioweapons prevention and defence, FY2001 through FY2007 (in $ millions)

federal funds allocated to bioweapons (Figure 5.3), and it is estimated that between 2001 and 2007, the US Government has allocated over $44 billion on biodefence and preparedness (Center for Arms Control and Non-Proliferation 2008).

A prominent project is BioShield, launched by the Bush Administration in 2004, and led by both the Department of Homeland Security (DHS) and the Department of Health and Human Services (HHS). With $5.6 billion to spend by 2014, it has no equivalent elsewhere in the world for its size. Project BioShield is a procurement programme, acting as a guarantee that the federal government will buy successfully-developed countermeasures for the Strategic National Stockpile (SNS).[16] In order to improve the performance of the project and to motivate more companies, the Bush Administration provides $1 billion as funding for research and development of vaccines and treatments under Project BioShield through a new regulatory division, the Biomedical Advanced Research and Development Authority (BARDA).

As for the issues addressed in this chapter, there is little doubt that a new set of technological dynamics has been created since the turn of the century. The project makes explicit what has long implicitly acknowledged, that is the total 'dual-use' potential of a large range of biotechnologies. In a conventional manner, in that technological field, 'dual-use' is defined as biological research with legitimate scientific purpose that may be misused to pose a biological threat to public health and/or national security (Department of Health and Human Services 2004, cited in Shea 2006). In reality, it might take on quite a different meaning from other industries, as

BOX 5.2 FOUR CLASSES OF MICROBIAL MANIPULATIONS IDENTIFIED BY NIXDORFF AND BENDER

Four classes of microbial manipulations that have been the subject of intense debate within and outside the scientific community are identified by Nixdorff and Bender:

1. The transfer of antibiotic resistance to micro organisms;
2. Modification of the antigenic properties of micro organisms;
3. Modification of the stability of micro-organisms to the environment;
   and
4. The transfer of pathogenic properties to micro-organisms.

Source: Nixdorff and Bender (2002)

the same organisms needed for the preparation of illness treatment may be at the same time part of the preparations for defensive and offensive use of biological weapons. Box 5.2 gives an example of the debate within the scientific community on the dual-use of microbes.

The dual-use potential of biotechnologies and genetic engineering are compounded by the coming of age of the security agenda, in which threats are not limited to the boundaries of the battlefield and coming from armies. In the security agenda identified in the 2002 *National Security Strategy of the United States of America*, and with not so many differences by the EU in its 2003 document, *European Security Strategy: A Secure Europe in a Better World*, threats come from the civilian side as well. Here, technology offers promising – and also ominous – perspectives. The emerging biotechnology of drug discovery promises great advances in medicine, biology, psychology, and a host of related sciences. This will boost the development of a range of agents that allow the manipulation of life, including the processes of cognition, development, reproduction and inheritance. Most industrialised countries have been increasing their research on these topics in recent years. The claimed objective is the development of more 'human', non-lethal weapons (also called 'less than lethal weapons') (Pearson and Malcolm 2003). The notion of non-lethal appears rather controversial because a categorical distinction between lethal and non-lethal chemical agents is not strictly possible, 'since "non-lethal" agents may be lethal at high concentration or for specific individuals. More seriously, synergy between two different non-lethal agents may make their combination highly lethal' (Wheelis 2002, p. 53).

What is even more worrying is the ambiguous international regulatory framework, because it is a driver of such 'non-lethal' research. The 1925 Geneva Protocol allowed for specific exemptions that permitted the use of 'riot control agents' (RCA) in domestic affairs. More recently, similar ambiguity persisted in the Chemical Weapons Convention, which despite explicitly prohibiting the use of RCA 'as a method of warfare', still includes some loopholes (*Chemical Weapons Convention Bulletin* 1994). Concern is that such agents could be used in interstate armed conflict and other military operations (counter-terrorism, peacekeeping, etc.).

It has been considered that the new biodefence research dollars represent a cornucopia for researchers in the biological sciences (Reppy 2004, p. 9). This statement is by no means groundless, as the largest component of the DHS S&T Directorate's budget is the Chemical and Biological Division, and in 2006, the DHS Research proper allocate 41 per cent of its budget to life sciences. According to the business association, Biotechnology Industry Organization (BIO), around 100 of the existing 1000 biotechnology companies are now involved in some form of biological defence research. For most of them, however, defence-related work comprises only a small part of their overall operations. R&D for biodefence seems more attractive to SMEs that find it difficult to raise capital through venture financing or the stock market, and are turning toward government biodefence contracts as a source of revenue rather than to large and highly capitalised companies (Shorett 2004). Not surprisingly, scientists running research institutes doing research in fields covered by BARDA argue that strong spin-off will stem from public DoD and DHS R&D.[17]

Reactions of business to the BioShield programme launched by the Bush Administration have been guarded so far.[18] Business asks to be protected against high-risk biomedical ventures that fail more often than traditional defence contractors (Smith et al. 2003). Strong lobbying by business may have succeeded in making the project more favourable to its demands. Some provisions introduced in BioShield II are business-friendly as they create 'competition-free' marketing incentives in patent regulation (up to the point of having no counterpart in international patent law; and therefore often perceived by other countries as illegally denying them inexpensive drugs), and tax incentives for investors and developers (Klotz 2006).

5. CONCLUSION

This chapter has argued against over-generalised claims on the relations between military and civilian technologies. Scholars of technical change generally adopt an institutionalist and evolutionary attitude which avoids making generalities of the kind mentioned in the introduction to this chapter. We propose to list five criteria on which any investigation of those relations should be based:

1. History A neoclassical approach, based on general equilibrium, is quite incapable of explaining the technological innovations and social changes of the past two centuries. Evolutionary economics has long observed the existence of cumulativeness and path dependence, with indeterminacy and openness as their corollary. It does not make much sense to compare quite different historical configurations of military–civilian technologies, e.g. poison arrows back to 1000 BC and the 21st century genetic sequence manipulation and its possible military and security uses. Even though it might resemble an Orwellian dream supported by enthusiastic analysts, it remains unchallenged that genetics offer quite radical innovations for the innermost reaches of the human body.[19]

Furthermore, another dimension of the historical process has to be investigated, one which looks at the complex interaction of technological innovations and structural change. History is not reducible to the arrow of time; it gets filled up with structures and institutions. It has been argued that a theoretical framework for a 'reasoned history', 'should investigate the social coordination of five subsystems, technological, scientific, economic, political, institutional and cultural' (Klotz 2006, p. 120).

2. International balance of power between countries This is probably one of the most underestimated drivers factored in by scholars of technical change. Of course, an approach based on national innovation systems (NIS) gives careful institutional substance to countries, all the more when the innovation system is defined in a broad sense as being rooted in the production and human resource development system and not only in the R&D system (Lundvall et al. 2002). This does not mean that nations are to be considered as a set of side-by-side stand-alone entities. This was already no longer the case by the end of the nineteenth century, when the 'first globalisation' era constituted the world economy as a whole, even though it was based on national competition and rivalry which led to colonial conquest and then to two world wars. It is still much less the case with contemporary globalisation, considered as an uneven systemic process (Serfati 2005). That the globalisation process does not discard the relevance of the NIS approach does not imply the reverse, that the impact of current globalisation on NIS can be abstracted from. The impact of globalisation is not only a reorganisation – albeit limited – of the hierarchy and balance of power *between* nations (erosion of American leadership, 'emerging countries', falling behind many LDC countries, etc.). The impact of globalisation also diffuses within countries, creating economic polarisation (to which the development of territorial clusters based on technological innovation and knowledge contributes) and increasing social inequalities.

In that context, it is (or should be) evident that relations between military and civilian technologies are not similar among countries. Apart from institutional variety the differences among countries appear closely connected to their (geo-economic) size and (international) power when defence issues, including relations between military and commercial technologies, are addressed (with some qualifications, see below). The development of the Internet, or the likely increase in military funding of R&D in biotechnology in the near future (on behalf of biodefence), reflects the unique place of the United States in the world. More generally the US defence R&D budget creates huge disproportions, even between nations who have also developed a technology policy based on defence R&D. This observation does not mean that only positive effects (spillovers) occurred in the USA. Instead, a number of scholars (Melman, Dumas) observed in the 1980s and 1990s the deterioration of US manufacturing competitiveness, in particular in sectors largely funded by the DoD (see section 1).

3. Sectors and technology Sectoral systems of innovation (SSI) (Malerba 2002) and technological systems (TS) (Carlsson and Stankiewicz 1991) have contributed to our understanding of technological change. In our view, these approaches should be seen as complementary to the NIS approach. As said by the initiators of this approach, a similar evolutionary and institutional theoretical framework is adopted. Those approaches could be useful to understand why the role of the military in different sectors is more or less important. They also constitute a good framework to analyse why, when the same level of military involvement occurs in different sectors, it falls short of producing similar results in terms of transfers of military to civilian technologies. Aeronautics is characterised by high returns to scale which legitimises state involvement, even for proponents of free trade (e.g. Paul Krugman (1991) and the 'new economics of geographic'). It is true that the handful of countries that developed an aeronautical industry, even in emerging countries (e.g. Brazil), have given a major role to the military, public procurement and protectionist measures. Returns to scale were not enough to justify military involvement; the role of aeronautics in military strategy was a factor at least as important as the former. A different situation applies in the pharmaceutical industry. In general, R&D public procurement has been significant in industrialised countries, but it has mainly operated through civilian ministerial bodies. Again, the regulatory framework could be quite different from one sector to another, bringing about differences in the potentiality of intersectoral transfer of technology, and also from the military side to the civilian one within the same sector. From a regulatory perspective, aeronautics, space and nuclear are different from electronics, machine-tools and biotechnology. The effects of the DoD's and DHS's recent involvement in biology occur in an industry that is R&D-intensive, but which does not necessarily require large-scale investment or specialized dedicated facilities as in aeronautics. Transfers between military and commercial applications are also different in this industry. Creating pathogenic weapons poses certain technical challenges, but the ability to produce enough material to cause morbidity, mortality, public panic and economic costs is within the capability of many laboratories (National Academy of Sciences 2004, p. 85), while mastering aerodynamics and rocket propulsion are more complicated.

4. Entrepreneurship Strong distortions created in the entrepreneurial culture of firms closely acquainted with ministries of defence have long been analysed as one of the major obstacles to the transformation of technologies designed for the military into products delivered to commercial markets. A reason put forward is that there is too high a wall of separation, a business culture on the military side that is ill-suited to engage in commercial production, and vice versa (Markusen and Yudken 1992). The culprit is military procurement, which has fostered inefficient production practices in the American economy (Melman 1970). What is striking is that similar difficulties or hesitation by defence contractors to transfer technologies designed for military purposes to commercial markets was observed in the USA, in France (Serfati 1995) and in the UK (Gummett and Stein 1997). Working with the military and having a large share of its turnover and profits dependant upon ministries of defence procurement clearly contributes to shaping an idiosyncratic entrepreneurial culture. Since diversification internal to the firm (from military to civilian activity) is a difficult process, the capacity of commercial-oriented manufacturing sectors to benefit from strong defence R&D depends, *inter alia*, on the 'breadth' of the industrial base, the number and the dynamism of commercial companies prepared to develop products derived from technologies and knowledge originating in the defence sector. On these grounds, some significant differences in the scope of intersectoral transfers between the USA and France can be accounted for by the breadth of the industrial base. In the USA, the existence of a large base of SMEs, along with financial markets and investors prepared to support them, facilitated such transfers. Instead, intersectoral transfers were rendered more difficult in France because of the longstanding weakness of entrepreneurial culture, coupled with the overarching role of large companies, mostly run by a management composed of personnel trained in leading technical (polytechnique) or administrative (Ecole Nationale d'Administration) institutes (Chesnais 1993).

5. Contingency There is no place here to engage in a long discussion on the relations between structural changes and contingency, which is a major epistemological problem.[20] The positivist motto gives evolution a single causal principle, which is closely associated with the myth of 'progress' in the sense of quest for perfection (or optimality for economists). Indeed, history moves on through multiple determinations and causalities, but there might be no logical contradistinction between recognising that some kinds of 'laws' govern economic and social relations and technological trajectories on the one hand, and the fact that contingency and 'unattended expectations' also produce 'history' on the other hand. The fact that technology is shaped by social needs, responds to economic incentives, and culminates with the accumulation of scientific knowledge, does not resolve in itself what Arthur calls the 'origination of radically novel technologies' (Arthur 2005, p. 16). The occurrence of a contingent or random event can produce bifurcation of trajectories which have nothing to do with a retrospective reconstruction based on the Panglossian view that evolution has resulted in the 'best of all possible worlds'. Here, evolutionary economists should carefully look at the fact that adaptation does not mean a monotonic (linear) evolution. The 9/11 terror attack was a singular event[21] which, not through itself but through the policy set in motion by the US Administration (a security agenda buttressed by a steep increase in the military budget), has opened a new stage in the relations between military and civilian technologies.

To sum up, it can be said that since the turn of the century a new era has commenced in the relations between military and civilian technologies through the security agenda. This new era is an outcome of combinations of the drivers proposed in this conclusion. They include a highly uneven and differentiating process of globalisation, the presence of deep-rooted and powerful military–industrial systems, the role of defence R&D funding in governmental policies since the end of World War II, and the unique configuration which allowed the USA to be the only 'superpower' (and spend over 50 per cent of world military expenditures and much more in defence R&D). The possibility offered separately or through cross-fertilisation by biotechnologies and nanotechnologies does not present only ominous prospects, it also deserves further attention from scholars of technical change.

NOTES

1. Despite its title, Ruttan's book (2006) is not on war, but offers a careful historical and geographical (US) enquiry into government subsidies and engagement in technological development. The obvious questions of why this accumulation of huge technological programmes has been carried out, and if the enforcement of such programmes does not inevitably lead to war, cannot be ignored. The author touches on that issue in the closing pages of his book, but does not give any clear answer.
2. And also the other way round, at least until the mid-19th century. Radical technologies developed from the 1850s onwards were opposed by the military. The Royal Navy resisted the introduction of steam and iron in the 1850s, and at the turn of the nineteenth century the internal combustion engine attracted very little interest from the military, as it developed extensively in aircraft and motor vehicles.
3. This chapter is focused on the relation between military and commercial technologies.
4. The term 'Big Science' first appeared in a 1961 article in *Science* magazine, entitled *Impact of Large-Scale Science on the United States*, by physicist and Oak Ridge National Laboratory director Alvin Weinberg. The article described Big Science as the new way of conducting science activities after World War II.
5. Texas Instruments, which demonstrated the first IC in 1958, had developed the device without DoD funding. Neither MOS nor microprocessor research (invented by Intel in 1972) was supported by the Pentagon (de Grasse 1984).
6. France ranks 11th in world production and 21st in exports of machine tools, and posted a huge deficit only exceeded among developed countries by the USA in 2005.
7. The economic definition in general is ambiguous, as it may refer either to 'any indirect effect of public expenditure' (*AnsMe Economic Dictionary* 2001) or to any non-market effect of economic activity, *private* or *public*.
8. See the presentation by William F. Slater III, Chicago Chapter of the Internet Society September 2002.
9. Other programmes, in particular nanotechnologies (€350 million) have military potential. Furthermore, the Joint Research Centre (JRC), the EU's Research Centre, originally established under the Euratom Treaty in 1957, carries out research in security-related fields. In September 2005, the JRC released a study, *Emerging Technologies in the Context of Security*, in which it describes security-related technologies as follows: 'Its key research focuses relate to integrated different technologies, interoperability and converging technologies. All other key technology reports are of high relevance to the security-related report: biotechnology, nanotechnology, research in the services sector, complexity and systemic, social sciences and humanities, cognitive science, agri-food and environmental technologies, energy technologies, ICT technologies, manufacturing technologies and transport-related research activities, as individual technologies but even more so as integrated' (JRC 2005, p. 41).
10. The military use of outer space comprises the whole range of space assets such as communications, Earth observation and navigation satellites.
11. Council Regulation no. 1321/2004 of 12 July 2004.
12. See Knight (2004). The EU wanted the standard, publicly available signal to use a modulation known as 'binary offset carrier' (BOC) 1.5. An alternative modulation, called BOC 1.1, has been suggested for Galileo by the USA.
13. 'Europe's rival to the US satellite positioning system does have a military role, the European Commission acknowledged on Wednesday, as a race to dominate space heated up', Andrew Bounds, 'Galileo to have military role', *Financial Times*, 16 May 2007.
14. 'The foundations of microbiology pioneered by Louis Pasteur and Robert Koch offered new prospects for those interested in biological weapons because it allowed agents to be chosen and designed on a rational basis' (Frischknecht 2003, S47).
15. Still, it does not proscribe acquisition or possession of chemical or biological materials.
16. The SNS contains pharmaceuticals, vaccines, medical supplies, and medical equipment to respond to terrorist attacks and other emergencies.

17. Anthony Fauci, head of the National Institute of Allergy and Infectious Diseases (NIAID), stated: 'Even if we never, ever get attacked, the benefits to society will be enormous', and 'it will undoubtedly have very positive spin-offs for things that have nothing to do with biodefence' (Motluk 2004).
18. See the statement of Peter Young, on behalf of BIO (Biotechnology Industry Organization), presented before the House Energy and Commerce Committee, Subcommittee on Health, 6 April 2006. The BIO represents more than 1100 biotechnology companies, academic institutions and state biotechnology centres.
19. See the research at DARPA whose objective is that, 'Future soldiers might have super-intelligence to go along with their super-endurance . . . "these super-soldiers may be able to go a week without sleeping if scientists can figure out how to copy the genetic wiring of whales and dolphins", because researchers are working to implant microchips into the brain to enhance cognitive performances' (Boot 2006, p. 454).
20. See Louçã (1997) for a discussion of these issues applied to long-term evolution of capitalism.
21. Even though it was not totally unpredictable, given the World Trade Center bombing on 26 February 1993.

REFERENCES

Abbate, J. (1999), *Inventing the Internet*, Cambridge, MA: MIT Press.

Arthur, W.B. (2005), 'The logic of invention', SFI working paper no. 05-12-045, accessed March 2007 at www.santafe.edu/research/publications/wpabstract/200512045.

Aspray, W. and O. Bernard (1994), 'Arming American scientists: NSF and the provision of scientific computing facilities for universities, 1950–1973', *IEEE Annals of the History of Computing*, **16** (4), 60–74.

Assembly of Western European Union (2006), The Interparliamentary European Security and Defence Assembly colloquy on space, defence and European security, 19–22 September, Kourou (French Guiana), accessed February 2008 at www.assembly.weu.int

Boot, M. (2006), *War Made New Technology, Warfare and the Course of History*, London: Gotham Books.

Burzykowska, A. (2006), 'ESDP and the space sector. Defining the architecture and mechanisms for effective cooperation', *Space Policy*, **22**, 35–41.

Carlsson, B. and R. Stankiewicz (1991), 'On the nature, function and composition of technological systems', *Journal of Evolutionary Economics*, **1**, 93–118.

Center for Arms Control and Non-Proliferation (2008), Federal funding for biological weapons prevention and defense, fiscal years 2001 to 2008, accessed February 2008 at www.armscontrolcenter.org/policy/biochem/fy08_biodefense_funding.

Cerf, Vinton G. (1993), 'How the Internet came to be, as told to Bernard Aboba', accessed September 2007 at www.virtualschool.edu/mon/Internet/CerfHowInternetCame2B.html.

Cerf, Vinton G. (1995), 'Computer networking, global infrastructure for the 21st century', accessed September 2007 at www.cs.washington.edu/homes/lazowska/cra/networks.html.

Chemical Weapons Convention (1994), 'Editorial: new technologies and the loophole in the convention', *CWC Bulletin*, **23**, 1–2.

Chesnais, F. (1993), 'The French national system of innovation', in R.R. Nelson (ed.), *National Innovation Systems: A Comparative Analysis*, New York: Oxford University Press.

Commission Communication (2004), 'On the implementation of the preparatory action on the enhancement of the European, industrial potential in the field of security research. Towards a programme to advance European security through research and technology' COM(2004) 72.

Commission Communication (2007), 'Galileo at a cross-road: the implementation of the European GNSS Programmes', to the European Parliament and the Council, Brussels, 16 May, COM(2007)261 final.

Cowan, R. and D. Foray (1995), 'Quandaries in the economics of dual technology and spillovers from military R&D', *Research Policy*, **24**, 851–68.

Dumas, L.J. (1986), *The Overburdened Economy*, Los Angeles, CA: University of California Press.

de Grasse R. (1984), 'The military and semiconductors', in J. Tirman (ed.), *The Militarization of High Technology*, Cambridge, MA: Ballinger Publishing, pp. 77–104.

Ergas, H. (1987), 'The importance of technology policy', in P. Dasgupta and P. Stoneman (eds), *Economic Policy and Technology Policy*, CEPR.

Flamm, K. (1998), 'Government's role in computers and superconductors', contract report for the Office of Technology Assessment, US Congress, Washington, DC.

Freeman, C. and L. Soete (1997), *The Economics of Industrial Innovation*, 3rd edn, Cambridge, MA: MIT Press.

Frischknecht, F. (2003), 'The history of biological warfare', *EMBO reports*, **4**, special issue.

Gansler, J.S. (1980), *The Defence Industry*, Cambridge, MA: MIT Press.

Geissler, E. and J.E. van Courtland Moon (eds) (1999), *Biological and Toxin Weapons: Research, Development and Use from the Middle Ages to 1945*, Chemical and biological warfare studies no. 18, London: Oxford University Press.

Gummett, P. and J.A. Stein (1997), 'Britain', in P. Gummett and J. Stein (eds), *European Defence Technology in Transition*, Amsterdam: Harwood Academic Publishers.

Hayes, B. (2006), 'Arming Big Brother. The EU's security research programme', TNI briefing series 1.

Hughes, T.P. (1998), *Rescuing Prometheus*, New York: Random House.

IEEE (n.d.), 'History of the technology', Chapter 4: 1972–1984, accessed February 2008 at www.ieee.org/web/aboutus/history_center/conferences/comsoc/chapter 4.html.

Joint Research Center (JRC) (2005), 'Emerging technologies in the context of "security" issues in the framework of science and technology', accessed September 2007 at http://serac.jrc.it/dmdocuments/emtec.pdf?PHPSESSID= a2f303179cff42d62d552634b9fd1ae3.

Klotz, L. (2006), 'BioShield marketing incentives to industry', Center for Arms Control and Non-Proliferation, January 18 electronic edition Center for Arms Control and Non-Proliferation CD72G06_603.

Knight, W. (2004), 'Row over GPS jamming still divides US and Europe', NewScientist.com news service, accessed December 2007 at www.newscientist.com/article.ns?id=dn4641.

Krugman, Paul (1991), 'Increasing returns and economic geography', *Journal of Political Economy*, **99** (3), 483–99.

Langlois, R.N. and D.C. Mowery (1996), 'The federal government role in the development of the US software industry' in D.C. Mowery (ed.), *The International Computer Software Industry: A Comparative Study of Industry Evolution and Structure*, New York: Oxford University Press.

Leitenberg, M. (2005), 'Assessing the biological weapons and bioterrorism threat', accessed December 2007 at www.armscontrolcenter.org/cbw/press/milton_assessing.shtml.

Louçã, F. (1997), *Turbulence in Economics: An Evolutionary Appraisal of Cycles and Complexity in Historical Processes*, Cheltenham, UK and Northampton, MA, USA: Edward Elgar.

Lundvall, B.Å, B. Johnson, S. Andersen, E. Sloth and B. Dalum (2002) 'National systems of production, innovation and competence building', *Research Policy*, **31**, 213–31.

Malerba, F. (2002), 'Sectoral systems of innovation and production'. *Research Policy*, **31**, 247–64.

Markusen, A. and J. Yudken (1992), *Dismantling the Cold War Economy*, New York: Basic Books.

Melman, Seymour (1969), *Pentagon Capitalism*, New York: McGraw-Hill.

Melman, S. (1970), *The Permanent War Economy*, New York: Simon and Schuster.

Molas-Gollart, J. (1997), 'Which way to go? Defence technology and the diversity of "dual-use" technology transfer', *Research Policy*, **26** (3), 367–85.

Motluck, Alison (2004), 'Protect and prosper', *New Scientist*, 23 October, accessed December 2007 at http://technology.newscientist.com/article/mg18424706.700-protect-and-prosper.html.

Mowery, D.C. and T.S. Simcoe (2002), 'Is the Internet a U.S. invention? An economic and technological history of computer networking', *Research Policy*, **31**, 1369–87.

National Academy of Sciences (2004), 'Biotechnology Research in an Age of Terrorism', accessed 23 February, 2007 at www.nap.edu/catalog/10827.html.

National Security Strategy of the United States of America (2002), accessed May 2003 at www://whitehouse.gov/ncs/nss.html.

Nixdorff, K. and W. Bender (2002), 'Biotechnology, ethics of research, and potential spin-off', *INESAP Information Bulletin*, http://www.inesap.org/bulletin19/bul19art05.htm.

OECD (2004), *The Security Economy*, Paris: OECD.

Pasco, X. (2006), 'A European approach to space security', Center for International and Security Studies at Maryland working paper, July, accessed September 2007 at www.cissm.umd.edu/papers/display.php?id=165.

Pearson, G.S. and D. Malcolm (2003), 'The danger to the Chemical Weapons Convention from incapacitating chemicals', March, series editors, Department of Peace Studies, University of Bradford, accessed 13 November, 2007 at http://microbiology.ucdavis.edu/faulty/mwheelis/cbw.html.

Reppy, J. (2004), 'Biotechnology research in an age of terrorism', The National Academy Press, accessed February 2008 at www.nap.edu/openbook.plip?ISBN=0309089778.

Research Group (2004), 'Research for a Secure Europe', report of the Group of Personalities in the Field of Security Research, European Commission, March accessed at http://europa.eu.int/comm/enterprise/security/doc/gop_en.pdf.

Ruttan, V.W. (2005), 'Military procurement and technology development', Staff paper series, March, accessed December 2007 at www.apec.umn.edu/faculty/vruttan/StaffPaperP05-3-MilitaryProcure.pdf.

Ruttan, V.W. (2006), *Is War Necessary for Economic Growth?*, Oxford: Oxford University Press.

Segal, B. (1995), 'A short history of Internet protocols at CERN', CERN IT-PDP-TE April, accessed February 2008 at http://ben.web.cern.ch/ben/TCPHIST.html.

Serfati, C. (1995), *Croissance, innovation et production d'armes*, Paris: Economica.

Serfati, C. (2004), *Impérialisme et militarisme. Actualité du vingt-et-unième siècle*, Lausanne: Editions Page2.

Serfati, C. (forthcoming), 'The role of defence innovation in national systems of innovation: some findings based on the French case', in A. James and P. Laredo (eds), *The Dynamics of Innovation in the Defence Sector: Economics, Technology and the New Security Environment*, Cheltenham UK and Northampton, MA, USA: Edward Elgar.

Shea, Dana A. (2006), 'Oversight of dual-use biological research: the National Science Advisory Board for Biosecurity', CRS Report for Congress, updated 10 July.

Shorett, P. (2004), 'The crack in bioshield's armor', *Genewatch*, **17** (5/6), 67–85.

Sidel, V. (2004), 'Bioshield, biosword', *Genewatch*, **17** (5/6), 33–45.

Slater, William F. III (2002) 'The history and growth of the Internet', 18 September, accessed February 2008 at www.isoc-chicago.org/highlights.html.

Smith, B.T., T.V. Inglesby and T. O'Toole (2003), 'Biodefense R&D: anticipating future threats, establishing a strategic environment', *Biosecurity & Bioterrorism*, **1** (3), 193–202.

Strategic Aerospace Review for the 21st century (STAR21) (2002), *Creating a Coherent Market and Policy Framework for a Vital European Industry*, July, Brussels: European Commission Enterprise Publications.

Stowsky, J. (2004), 'Secrets to shield or share? New dilemmas for military R&D policy in the digital age', *Research Policy*, **33**, 1369–87.

US Department of Labor (2007), 'Homeland security', accessed September at www.doleta.gov/business/.

Wheelis, M. (2002), 'Biotechnology and biochemical weapons', *Nonproliferation Review*, **9** (1), 48–53.

World Economic Forum (2007), 'Global risks 2007', Global Risk Network Report accessed February 2008 at www.useforum.org/pdf/CSI/Global_Risks_2007.pdf.

6. Theory and practice in knowledge transfer: the emergence of 'interface structures'

Elena Castro-Martínez, Ignacio Fernández-de-Lucio and Jordi Molas-Gallart

1. BACKGROUND: DEFINING AND CHARACTERISING INTERFACE STRUCTURES

The development of innovation systems approaches is closely connected with policy practice. Policy agencies in Scandinavian countries played an important role in their early development (Carlsson 2002). Later, the OECD contributed to the development of the approach and its diffusion among academics and policy-makers (Godin 2004). Our chapter is also rooted in practice. Based on the applied experience of existing organisations, we will define and characterise a new type of structure within an innovation system.

Innovation systems approaches focus the attention of policy-makers on the existence of situations where the relationships between different actors in the system are weak, and where divergent cultural traits and organisational practices pose a barrier to the collaboration and coordination among different actors in a system (Polt et al. 2001). There has been abundant research studying these relationships, especially in areas like university–industry linkages where cultural differences and other barriers have hampered a fluid relationship between the universities and their economic environment (Agrawal 2001; Carlsson et al. 2002; Carayol 2003).

This literature is often normative, proposing policy instruments to overcome barriers and strengthen the interconnectedness of the system. Yet it usually takes as its subject the problems emerging in countries where there is both an advanced research and industrial infrastructure. Here cultural differences, managerial difficulties, and divergent interests, all pose barriers to technology transfer and other forms of relationship, but there is an assumption that, in the absence of these barriers, potential technology customers would be able to articulate a demand that could then be serviced by public research organisations. If some difficulties are overcome and enough information is available then relationships will develop.

The problems we address in this chapter are even more complex. We focus on situations in which, in addition, there is a lack of awareness of what oneself and the other relevant actors in the system can offer or require. In other words, we address parts of the system in which the actors 'don't know what they don't know'. To respond to this situation, policy practitioners in different contexts have devised remarkably similar responses, which have an organisational rather than a purely instrumental nature. They have developed what we have defined elsewhere as 'interface structures' (Fernández-de-Lucio et al. 1996). This chapter develops this notion.

The objective of an interface structure is to act as a promoter and catalyst of the relationships between different actors in the system, reducing the barriers that have made it difficult for them to establish long-term relationships, but without becoming directly involved in the innovation process itself. The chapter will show that these interface structures have much broader roles than those typically attributed to the much studied technology transfer organisations.

Based on case studies of two different experiences we define the concept of 'interface structure', identify its role and characteristics (networked structure, broad portfolio of activities, etc.), and support for the relationships among different system actors. The chapter presents the case studies of a network of university–industry technology transfer organisations set up in Spain in the late 1980s (OTRI Network), and an organisation set up in the United Kingdom to improve the relationship between defence research capacities and needs and their commercial environment (the Defence Diversification Agency). After the case studies we map out some common traits that bring together these initiatives originating in such different contexts, and based on these traits we develop the concept of the interface structure. We conclude by placing this context within a systemic view of innovation.

2. THE OTRI NETWORK AND UNIVERSITY–INDUSTRY TECHNOLOGY TRANSFER IN SPAIN

2.1 Background

During the mid-1980s the Spanish innovation system was being rapidly transformed. Spain was emerging from almost 40 years of military dictatorship (a new democratic constitution was approved in 1978) with a model of economic development that has sometimes been described as 'dependent': research was virtually non-existent, and the little that was done was often carried out in large government research organisations with weak linkages to their socio-economic environment. In the mid-1980s a whole set of legal initiatives was enacted attempting to develop new research and innovation capacities (University Reform Act – 1983, Act for the Promotion and Coordination of Scientific and Technical Research, usually referred to as the Science Act – 1986, and a new Patent Act – 1986). Linked to them the First National R&D plan was launched in 1988. Until then, universities had carried out very little research, typically funded through their own resources. Until the University Reform Act, it was also illegal for universities to obtain resources additional to their normal budgets by entering, for instance, into contractual agreements with firms and other organisations. The only linkage between universities and firms was organised through four 'University–Enterprise Foundations' that had been created during the 1970s and early 1980s, to manage the contracts that research groups from a few universities might have obtained through their own initiative.

The situation in the government research organisations had been marginally better. The Spanish Council for Scientific Research (Consejo Superior de Investigaciones Científicas – CSIC), the main research organisation in the country, had several applied research institutes with a tradition of collaboration with industry, although paradoxically, this had weakened substantially since the 1950s (López García 1999). In the mid-1980s CSIC launched several initiatives to promote collaboration with industry, like the creation of a technology transfer office (1985), and the launch of cooperative programme (PECOS) (Fernández-de-Lucio 1988).

On the demand side, the Spanish development model had been based on cheap labour and energy costs, foreign investment and foreign technology (García Delgado 1995). During the 1960s and 1970s, Spanish firms carried out virtually no R&D: in 1964 Spanish firms spent 0.03 per cent of GDP, a percentage that, by 1984 had grown to a less-than-modest 0.18 per cent. The lack of formal research activities was accompanied by the low educational level of the Spanish industrial labour force: by 1984 only 2 per cent of industrial employees had a university degree (Más et al. 2004).

Within this context, policy-makers were aware of the need to support the collaboration between public research organisations and industry. The managers of the National Research Plan were asked to support a relationship that had been, until then, non-existent. Rather than supporting university–industry collaboration, science policy managers were given the difficult job of creating it.

2.2 The OTRI Network: Genesis and Main Characteristics

The creation of technology transfer offices in Spanish universities was a 'top-down' policy initiative. In 1989 the managers of the National R&D Plan included among their activities the setting up of Oficinas de Transferencia de Resultados de la Investigación (Offices for the Transfer of Research Results, OTRIs) in all Spanish universities. The OTRIs were originally designed with the following objectives, among others (CICYT 1988):

- To identify capacities and results with potential non-academic applications generated by research groups;
- To facilitate the transfer of research results from the research groups to industry and other users;
- To support researchers when entering contractual negotiations with firms and other clients and help them manage research contracts and protect their intellectual property with the appropriate legal tools;
- To provide researchers with information on opportunities available through European programmes and help them develop research projects.

These were all areas in which Spanish universities and research centres had little if any experience. The OTRIs' main goal was to facilitate university–industry collaboration, but they had a broad set of objectives. For instance the managers of the Spanish National R&D Plan used the OTRIs as conveyors of information to the academic community, and relied on them to disseminate information and support access to the different policy instruments in support of university–industry relations that they would develop and implement over the years.

Importantly, the OTRIs were set up as a network co-ordinated and supported by a central unit (the Office of Technology Transfer – OTT) within the National Plan Secretariat (Castro-Martínez and de-Lucio 1991; Fernández-de-Lucio et al. 1996).[1] The role of this central co-ordination was paramount. The OTRIs themselves were very small units, intended to be close to their local research capacities (mainly in the universities) and local needs. Yet, for this same reason, they lacked specialised capacities and, importantly, experience. The central OTT coordinated activities, facilitated the exchange of information and experiences, and provided, among other services, the dissemination of information on research opportunities and management training in areas like intellectual property.

As the OTRIs themselves gained capacity and experience the central support role of the OTT diminished in importance, and the Office was formally closed in the late 1990s. However, the OTRIs continued to work as a network. In 1997 the Vice-Chancellors of all Spanish universities formally created the OTRI Network, which continues to organise training, establish working groups, and develop other activities typically oriented towards the dissemination of good practice (www.redotriuniversidades.net).

All Spanish universities, and the CSIC, joined the initiative and set up their OTRIs. Similarly, all newly created universities have, since their establishment, set up OTRIs affiliated to the network. The outcome of this process was the establishment of a geographically distributed network of small units, displaying close links with and detailed knowledge of their local environments.[2]

The emergence of a network of small units was not an accident, but was designed into the policy initiative. An OTRI would have two or three employees, who would operate as generalists without a clear definition of their functions, supported by an administrative post. Their main role was to stimulate the relationships between industry and academia, on the one hand encouraging academic researchers to get involved in work with industrial partners by offering information and personalised support, while on the other hand, providing information to the industrial community about the capabilities available in academic research centres and universities. In this way, the OTRI played the role of a facilitator, rather than that of a technology broker, trying to support the creation of a market for research services and outputs where none existed. Spanish academics had developed an inward-looking culture, oriented towards teaching, less concerned by research, and without any interest in technology transfer activities. Within this context, the OTRI had to operate as a client-oriented service, avoiding the bureaucratic and legalistic approaches that are still today characteristic of universities' and research organisations' administrations.

Given the distributed nature of the OTRI network, not all OTRIs developed following exactly the same model. Some developed more rapidly, particularly those situated within universities with applied research capacities, and where groups of lecturers became rapidly involved and increasingly experienced in establishing contractual relationships with the private sector. Often, OTRIs had to take over managerial functions for which they had not been originally intended, as the central university services proved unable to administer research contracts and programmes. In these cases the OTRIs became overwhelmed with administrative tasks and found it difficult to engage in the functions for which they had been originally intended.

The OTRIs were initially funded by central government (through the Spanish National R&D Plan), but progressively the universities shouldered a growing share of core funding (personnel, equipment, etc.). They remain, however, dependent on public support (both national and regional) to develop their activities. Currently 35 per cent of their funding comes from public sector grants (RedOTRI 2006).

2.3 Outcomes

Whatever the difficulties experienced by specific OTRIs, there can be little doubt that they play an important role in the major change that Spanish universities were to experience in the late 1980s and early 1990s. For instance, according to OECD data, the share of universities' R&D funded by firms grew from 0.4 per cent in 1981 to 7.5 per cent in 2005, higher than the OECD average (OECD 2004). This might be attributed to causes other than the emergence of these offices; yet Table 6.1 shows how several indicators of OTRI activity and outcomes have grown rapidly, suggesting an important role in drawing public research funds to universities. Even when the main goal of the offices was not the generation of research contracts but to establish links and generate mutual interest between two hitherto separated communities, they managed an increasingly large number of contracts that generated substantial resources.

The amount and value of R&D contracts managed have recently stabilised after a protracted period of rapid growth, reflecting probably that the community of university researchers that were receptive to the OTRIs' 'message' and capable of engaging in contractual research relationships with external partners has been reached in its capacity, and that the growth generated by the creation of new universities and the hiring of young academics has stopped. Recently the OTRIs have been involved in supporting new transfer tools like the creation of spin-off firms in science-based sectors. This is a new avenue likely to require further cultural, managerial and legal changes.

3. THE UK DEFENCE DIVERSIFICATION AGENCY AND MILITARY–CIVILIAN TECHNOLOGY TRANSFER

3.1 Background: Commercialisation and the Government Defence Labs

The creation of the UK Defence Diversification Agency (DDA) is linked to the process of commercialisation in UK defence research establishments, which is in turn part of a series of reforms in government research establishments implemented mainly during the 1980s and 1990s. Policy measures seeking to obtain more 'value-for-money' from government research investments introduced greater accountability and a growing commercialisation of the relationship with government users. In the defence area, most of the previously independent establishments were progressively merged into the Defence Research Agency (1991) and the Defence Evaluation and Research Agency – DERA (1995), which committed to seeking maximum

Table 6.1 University OTRIs: activity and output indicators

	1989	1995	2001	2002	2003	2004	2005
Number of universities	32	48	57	57	58	58	60
Professional staff (FTE)	64	154	179	234	294	348	434
Number of R&D contracts	695	3270	8687	8323	7958	9830	10 088
Income from R&D contracts (million €)	8	120	218	252	258	282	339
Number of national patent applications	24	140	264	301	317	307	336
Number of international patent extensions			43	66	107	93	117
Number of options and licence contracts			50	53	78	143	106
Income from licences (million €)			0.5	1.1	1.7	1.9	1.7
New spin-offs created			39	65	87	90	88

Sources: Own elaboration of CICYT annual reports (1989 and 1995) and University's OTRI Network annual surveys (2001–2005)

wealth creation from its research activities. As executive agencies of the Ministry of Defence (MoD), these organisations gained the flexibility to become more commercial in their operations resulting in significant growth in DERA's external income and the development of a more entrepreneurial culture among its scientific and technical staff (Arthur D. Little 2000).[3] Later most of DERA was to turn into a new firm (QinetiQ) which was in the process of being floated on the stock market at the time of writing.

The creation of the DDA is related to this process. The Labour government set up the DDA to address three priorities:

- To encourage the widest possible exploitation of military technology by civilian firms;
- To encourage a growing variety of defence suppliers;
- To encourage the transfer of suitable civil technology into military programmes.

To these ends the DDA was tasked with encouraging access to government defence labs, providing information about the capabilities they had available, stimulating transfer of the MoD's intellectual property rights, seeking partnership with companies for programmes of co-development and adaptation, and providing information about future equipment needs and technological trends to potential new suppliers. Therefore, the objectives laid out for the DDA considered the transfer of technology from military to civilian applications, and vice versa. Initially, the model implicit in the description of DDA's tasks was linear: potential technology users in the civilian industry would be made aware of the technology offerings from DERA, or of opportunities in the defence markets. The possibility of more interactive collaboration through the engagement in programmes of co-development and technology adaptation was mentioned but it did not feature prominently; instead there was a focus on providing information through, for instance, databases of technological capabilities to which it was expected civilian firms would respond. The proposal could be seen as a continuation of other attempts at organising and stimulating technology transfer from the defence research establishments that had been tried with little success during the 1980s and early 1990s.[4] As we will see below the model that was eventually implemented diverted from this initial sketch.

3.2 Implementation: the DDA

The Defence Diversification Agency (DDA) was established in 1998 and began its first year of operation with a budget of £2 million. During its first year of operation the DDA created a network of support activities and personnel across the country. It was set up with a small centre and, initially, five regional offices. With an original remit of promoting technology spillover between the defence and civilian sectors, the challenge facing the fledgling organisation was to establish mechanisms able to deal with a large variety of technological fields and industrial sectors. The defence research establishments had developed technological capabilities across a broad range of generic technologies with potential applications in many sectors, and the range of civilian technologies applicable to defence systems were not limited to any specific sector. Unlike other technology transfer organisations, which typically focused on a relatively narrow range of technologies and sectors, the DDA had to develop activities covering virtually all technological sectors.

Further there was no tradition among civilian firms of access to defence research capacities. The latter had developed in a closed environment, and there was a widespread belief outside defence circles, that their activities were very specialised and irrelevant to civilian needs. Further, the highly publicised and recurrent scandals in the industry involving time and cost overruns, and dubious export practices involving the payment of bribes and other 'sweeteners' to potential customers, had built an image of the defence industries as inefficient and stuck to outmoded managerial practices.

This situation created substantial challenges. First, there was not a market 'out there' waiting to access the capacities that the defence labs had to offer;

on the contrary a customer base had to be built virtually from scratch. To do so, the DDA established a network of regional representatives (Technology Diversification Managers – TDMs) working in close collaboration with the local Regional Development Authorities,[5] and other local organisations including the government-funded 'Business Links'. By 2004 the DDA was working from 26 different locations, each with at least one TDM. The TDM's objective was to be close to the local needs and to establish links with firms and organisations that could, in principle, be interested in working with the defence labs or for the defence agencies.

The second problem was how to identify potential partners operating in different sectors with very limited availability of resources. Often, the technological requirements of a firm would not have been articulated; how could one then identify possible fields for technological co-operation when it was not known what a potential partner would need? To address this problem, the DDA developed a portfolio of tools to carry out technology audits of firms, and identify possible technology requirements. The TDMs apply these tools free of charge to selected firms. The Regional Development Authorities and the TDMs identify firms that, for their characteristics, may benefit from the technology offerings available at the defence labs, and offer them a technology audit for free. If firms want to pursue the opportunities identified through the audit, the DDA will facilitate contacts and a further exploratory analysis with scientists and technicians from the defence research establishments. The DDA will usually step aside from the process once these contacts have been established and the future partners start moving towards the signature of contractual agreements.

The main DDA stakeholder, the UK Ministry of Defence, soon focused its attention on another of the DDA's original objectives: the identification of civilian technologies of interest to military applications. Driven by the growing importance of civilian-led IT for military applications, the problem here has not only been how to find relevant technological capabilities among civilian suppliers, but how to insert them in time into new weapons systems. The distributed network of TDMs has also helped in the identification of local capabilities that could be of interest to defence production.

3.3 Outcomes

Since its creation the DDA has grown to reach a total staff of 60, distributed between the central offices and a regional network of 26 TDMs. It has generated a large number of collaborations between firms and the defence research laboratories, and has introduced many commercial firms to defence markets. Since its establishment the DDA has advised technology transfer contracts with a direct value of £17 million. In 2004, for instance, it placed 18 contracts between firms that had so far focused on civilian markets, and the Ministry of Defence, and helped broker 104 technology transfer contracts. Yet, it would be misleading to assess the outcomes of the DDA's work by using the type of indicators that have been applied to traditional technology brokers. Brokering technology transfer deals is not the focus of DDA's activity. The flexible approach to the identification of technological opportunities implemented by the DDA has led to a broad variety of initiatives including, among others:

Technology and knowledge brokering Matching the technological needs of commercial firms with the capabilities existing at the defence research laboratories. At times this will result in the licensing of specific technologies, but more often the defence research establishments will provide technology services, mainly the testing and evaluation of new technologies and products. This activity revolves around the implementation of technology audits in SMEs. Currently the DDA carries out about 160 audits each year.

Marketing assistance Identifying potential new suppliers of advanced technologies to the UK Ministry of Defence.

Regeneration activities The DDA has, for instance, contributed to the development of an under-used military airfield into a specialist test and evaluation centre for both military and civilian unmanned aerial vehicles.

Managing incubators The Farnborough Enterprise Hub and the London Business Innovation Centre are two examples of incubators managed by the DDA and supported by a group of regional and local authorities

Dissemination Participating in and organising events to present networking opportunities, introduce the procurement needs and practices of the MoD, discuss financing mechanisms for new companies and products, and explain intellectual property protection and evaluation practices. These activities are typically oriented to SMEs.

Development and implementation of knowledge and technology management tools In addition to its own technology audit methodology, the DDA is developing, together with a British SME, a software tool to quantify the effect of technology and knowledge transfer.

Project evaluation The DDA has helped the Welsh Knowledge Exploitation Fund develop a process for the evaluation of research proposals, identifying scientists from defence laboratories able to provide technical and scientific appraisals of the proposals.

This is a broad set of activities spreading over a broad variety of industrial sectors, technologies, and regions. Examples of technologies that, having been first developed for a military use, have been developed into civilian applications with the help of the DDA cover virtually all industrial sectors, including electronics, health, renewable energies, transport, industrial coatings, monitoring services, etc.

4. ANALYSIS: TOWARDS 'INTERFACE STRUCTURES'

Although they emerge from very different contexts and address different communities, the OTRI Network and the DDA experience share interesting commonalities. In both cases the problem they faced was to create a relationship between actors that did not have a tradition of collaboration. In Spain, there had hardly been any relationship between universities and industry prior to the creation of the OTRI Network. In the UK, civilian firms, particularly SMEs, had no experience of collaborating with defence research laboratories. In both cases, the parties were unaware of each other's capabilities and needs, and no 'market' for technology or technology services existed. It is our hypothesis that under these circumstances traditional 'technology broker' functions oriented towards technology commercialisation will not be adequate.

What is remarkable about the experience of the OTRI Network and the DDA is that without any previous knowledge of each other they developed a similar approach, which translates into a different kind of organisation we call an 'interface structure'. We use this term to refer to organisations set up by one or more agents to promote and facilitate relationships in all matters related to innovation between different actors of an innovation system (Fernández-de-Lucio et al. 1996). Interface structures have a set of distinctive and common characteristics:

'Catalyst role' In chemistry a catalyst helps two other elements to interact in a chemical reaction, which would not occur in the absence of the catalyst. The catalyst however does not participate in the reaction, thus a good catalyst is one that does not spend itself in it. This is a good analogy with the role of the 'interface structures' we are analysing here. Their role is to put in touch two actors that would not otherwise interact, and to assist in reaching any agreement to take the relationship further. Interface structures inject a new 'dynamic' into the innovation system, making its actors

increasingly aware of opportunities in other parts of the system, and more active through the establishment of new relationships, without the interface structure itself getting directly involved in these new relations. The OTRIs may participate in licensing and other contractual agreements, but only as a further service they offer to their 'customers' in universities and research organisations. Their strategy can be considered 'mixed', having both commercialisation and catalyst roles, but with the latter usually dominating. The DDA case is even clearer in this respect. It does not get involved in technology commercialisation activities: its practice is to step aside once two groups start 'talking business'. In this respect, it cannot be more different from a technology broker, which derives its income and main 'raison d'être' from licensing deals and other commercialisation practices.

Decentralised and networked To be able to provide the networking opportunities and act as a catalyst, interface structures need to be close to their potential clients and, therefore, have to establish a network of geographically dispersed operations. Both the OTRI Network and the DDA have offices all over the national geography, and have a system to network their operations. A network is necessary because of the small size of the local offices (a few individuals at most, sometimes only one person).

Customer-led Interface structures are not clearing houses for technologies and capabilities developed in research organisations; in other words their role is *not* to push technologies developed by scientific organisation to new sectors and customers. Instead, one of the problems they face is that the communities with which researchers could establish mutually profitable links have not developed and articulated their demand for technology. At the same time the scientists and technicians are not aware of what they could offer potential commercial customers. The orientation of the interface structure is therefore customer-led: helping companies define and convey their technological needs, and helping researchers understand better how their capabilities can be used outside their normal practice environment (in our cases academic or defence). The DDA, for instance, stimulates technological demand through the use of formal analytical tools (their technology audits). The OTRI Network has focused on advising and training researchers in the management and commercial practices necessary to be able to operate outside the academic environment.

Individualised approach and 'know-who' capacities To be able to establish and strengthen networks at the local and regional level, interface structures have to emphasise personal contacts, and be able to respond to the individual demands of specific researchers and potential users/beneficiaries of their technological capabilities. For instance, while one of the DDA's failed predecessors, Defence Technologies Enterprises Ltd, built a database of technologies on offer, the DDA is building a database of thousands of contacts in industry and academia, seeing its growing network of contacts as one of its main assets. In Spain, although the OTRI Network/OTT did not build a centralised contact database, it did focus a substantial part of its activity on providing points of contact to both industrialists and researchers. In practice both organisations develop a 'know-who' expertise (OECD 1996) as a crucial capacity to support their daily activities.

Generalist rather than specialised Traditional technology brokers that operate as commercial intermediaries and seek to generate funds from the licensing of technologies and other commercial activities are typically targeted in their operations and the set of skills they have to deploy, often focusing on a few technological areas. In contrast, interface structures have a broader set of goals and activities, will have to rely on generalists who are able to generate new personal networks, and will not focus on narrow disciplinary or technological specialisations. The networked and customer-led characteristics of interface structures cause them to be involved in a variety of transfer and collaborative mechanisms, and of sectors of application and technology fields. These call for a flexible technological and sectoral coverage to be able to adapt to the needs of the client base, requiring their local representatives to be generalists able to network within and across different communities ('know-who'), rather than technical or functional specialists. Functional expertise (for instance on legal or financial matters) will be provided through the central services (DDA), or through the networking activities (OTRI Network).

5. CONCLUSIONS

As we have seen in this chapter, interface structures can emerge in very different contexts, but respond to a similar problem: the need to bridge different communities and institutions, with different cultures and practices, in which both sides are unable to articulate their technological needs or offerings, and unaware of the capabilities or needs the other might have. From this perspective the concept of interface structures can be integrated into a 'systems of innovation' framework. The problem that interface structures address is the weakness of the linkages between system components; in other words, instead of operating over the components of an innovation system, the interface structures facilitate the linkages among them. They are therefore relevant in situations in which the relationships between elements of the innovation system are not occurring spontaneously or fluidly; that is, in systems that lack 'articulation' (Fernández-de-Lucio et al. 2003).

The roles of an interface structure are complex and go far beyond the development of an organisation to operate as a conveyor belt for technologies. Their positions within a system of innovation will, however, tend to be weak. Without a clear commercial orientation, interface structures require the support of other organisations, which in the cases we have analysed are always located within the public sector. The DDA depends on the budgetary support of the Ministry of Defence and relies on the assistance of Regional Development Authorities. The OTRI Network was initially funded by central government (through the Spanish National R&D Plan) and regional authorities and even now depends too much on national or regional subsidies (35 per cent of their funding). Within this context both organisations can easily be affected by policy changes. Pressures to become, for instance, more commercial in their strategies and day-to-day practices would completely change their nature, affecting most if not all of the characteristics we have identified in this chapter.[6] These pressures are likely to develop, particularly as the impacts of interface structures are often difficult to translate into easy-to-interpret quantitative indicators and may appear to be vague for policy-makers. How does one for instance, translate the 'strengthening of the linkages in an innovation system' into a clear measure of policy impact on social and economic welfare? This difficulty of presenting the rationale for interface structures through 'hard' impact data and using concepts that do not rest on complex theoretical frameworks (innovation systems) possibly constitutes their main weakness today as a policy instrument, and explains why their diffusion is still limited.

NOTES

1. The arrangement was called OTRI Network/OTT.
2. Our previous research has shown how, in fact, all OTRIs have established collaborative links with firms from their specific regions (Castro-Martínez et al. 2005).
3. However, DERA's main function remained to support the needs of the Ministry of Defence.
4. For instance, on the failed experience in the late 1980s with the creation of an independent technology broker to commercialise defence technology (Defence Technology Enterprises) see Spinardi (1992). For a detailed analysis of another, more successful, experience with more sophisticated models of technology transfer, see the analysis by Molas-Gallart and Sinclair (1999) of the Dual-Use Technology Centres.
5. The Regional Development Authorities are government-funded agencies whose main task is the promotion of regional economic development.
6. The OTRIs have been pushed to diversify their sources of income by increasingly focusing on project-based funding.

REFERENCES

Agrawal, A. (2001), 'University-to-industry knowledge transfer literature review and unanswered questions', *International Journal of Management Reviews*, **3**, 285–302.

Arthur D. Little (2000), *Value from Research: Achieving Innovation with LPRIs*, Brussels and Luxembourg: ECSC-EC-EAEC.

Carlsson, B. (2002), 'Preface', in Bo Carlsson (ed.), *Technological Systems in the Bio Industries: An International Study*, Norwell, USA: Kluwer Academic Publishers, pp. vii–viii.

Carlsson, B., S. Jacobsson, M. Holmén and A. Rickne (2002), 'Innovation systems: analytical and methodological issues', *Research Policy*, **31**, 233–45.

Carayol, N. (2003), 'Objectives, agreements and matching in science–industry collaborations: reassembling the pieces of the puzzle', *Research Policy*, **32** (6), 887–908.

Castro-Martínez, E. and I. Fernández-de-Lucio (1991), 'Hacia un Sistema integrado Ciencia-Tecnología-Industria', in Maria Angeles Roque (ed.) *Noves tecnologies i desafiament sòcio-econòmic*, Barcelona: Institut Catalá d'Estudis Mediterranis, pp. 315–24.

Castro-Martínez, E., A.M. Cortes-Herreros, M. Nicolás-Gelonch and C. Costa-Leja (2005), 'Una aproximación al análisis del impacto de las universidades en su entorno a través de un estudio de las actividades de las OTRI universitarias españolas', in 'Proceedings XI Seminario Latino-Iberoamericano de Gestión Tecnológica (ALTEC 2005)', México.

CICYT (1988), 'Las Oficinas de Transferencia de resultados de Investigación y la Oficina de Transferencia de Tecnología', CICYT internal memo, Madrid.

Fernández-de-Lucio, I. (1988), 'Valoración y transferencia de tecnología', CSIC internal memo, Madrid.

Fernández-de-Lucio, I., F. Conesa, M. Garea, E. Castro, A. Gutiérrez and M.A. Bodegas (1996), *Estructuras de interfaz en el Sistema español de Innovación. Su papel en la difusión de tecnología*, Valencia: Universidad Politécnica de Valencia.

Fernández-de-Lucio, I., F. Jiménez-Sáez, A. Gutiérrez-Gracia and E. Castro-Martínez (2003), 'The evaluation of the national food technology programme: the articulation of the Spanish food innovation system', *International Journal Agricultural Resources, Governance and Ecology*, **2** (3/4), 343–60.

García Delgado, J.L. (1995), 'La economía española durante el franquismo', *Revista Temas*, **12**, 27–32.

Godin, B. (2004), 'The New Economy: what the concept owes to the OECD', *Research Policy*, **33** (5), 679–90.

Grovlen, M. and K. Aarvak (1997), 'Technology transfer from R&D institutions to SMEs. The Norwegian TEFT Programme', *Industry & Higher Education*, **11** (1), 50–3.

López García, S. (1999), 'El Patronato Juan de la Cierva (1939-1960), III Parte: La investigación científica y tecnológica', *Arbor*, **637** (CLXII), 1–32.

Mas, M., F. Pérez, E. Uriel, L. Serrano and A. Soler (2004), 'Capital Humano en España y su distribución provincial. Banco de datos', IVIE (Valencia), accessed 12 March 2007 at www.ivie.es/banco/capital.php.

Molas-Gallart, J. and T. Sinclair (1999), 'From technology generation to technology transfer: the concept and reality of the Dual-Use Technology Centres', *Technovation*, **11**, 661–71.

Organisation for Economic Development and Co-operation (OECD) (1996), 'The knowledge-based economy', reference no. OCDE/GD (96)102, Paris.

OECD (2004), Main science and technology indicators, Paris: OECD.

Polt, W., C. Rammer, D. Scharting, H. Gassler and A. Schibany (2001), 'Benchmarking industry–science relations: the role of framework conditions', *Science and Public Policy*, **28** (4), 247–58.

RedOTRI Universidades (2006), 'Informe RedOTRI 2006', www.redotriuniversidades.net, acceded 12 March 2007.

Spinardi, G. (1992), 'Defence technology enterprises', *Science and Public Policy*, **19** (4), 198–206.

7. Dual technological knowledge and the firm's trade-off between civilian and military activities

Michel Callois

1. INTRODUCTION

Since the seminal work of Lundvall (1992), knowledge has been considered as an asset that firms must be able to manage in order to gain competitive advantage. As an asset, knowledge is a central feature of a firm's production in several ways: research and development, cooperation in innovation through networks, social relationship between members inside the organization, etc. Knowledge presents some characteristics of a public good. In other words, it cannot be easily transferred between two entities, nor can it be 'fully' appropriated by the firms that have produced it. In fact, this approach reveals the determinant that modifies its production, its significance and its mobility. In this perspective, knowledge can be observed as a system where forces like firms' behaviours, institutions, and inter- and intra-industry relationships tend to shape its economic meaning.

The starting point of this chapter is to analyse the case of the defence industry as a specific institutional environment. During the Cold War, the defence industry was based on a predominant single user environment, whose primary goal was to prevent threats in the short and long run. Thus, government-based demand shaped the technological opportunities of firms in order to achieve this critical goal. During the post Cold War era, this 'programme-based' role switched to a 'client-based' role. This had a deep impact on the behaviour of defence competitors: the reconfiguration of the aerospace industry in the USA and Europe is a straightforward example. Some key concepts were developed to explain the difficult dilemma faced by defence firms: should defence firms define their knowledge strategy on the defence industry exclusively, or civilian activities (Kulve and Smit 2003)? Is the knowledge acquired by defence firms suitable for civilian markets? Is defence-related knowledge subject to secrecy (Graham 2004)?

In our understanding of the importance of this trade-off, the notions of systems of innovation, both sectoral (Malerba 2002) and national (Lundvall 1992), highlight an important point: innovation is not an isolated activity in economic life. In fact, innovation cannot be studied without analysing the context where it takes place: 'a sectoral system of innovation and production is a set of new and established products for specific use and the set of agents carrying out market and non-market interactions for the creation, production and sale of those products. Sectoral systems have a knowledge base, technologies, inputs and demand' (Malerba 2002, p. 250).

Thus, there is a strong link between three dimensions:

- Specificity of the space where the firm acts (institutional framework);
- The relationship between actors (connectedness);
- The knowledge base of firms (cognitive).

Researchers have pointed out the importance of networks as a key tool to drive innovation. The idea is that knowledge has some particular specificity concerning its creation, its appropriation and its transfer. As Jensen et al. (2004) stressed: 'One way to indicate that the transfer of knowledge is a demanding process has been to point to the need for establishing "absorptive capacity". . . . This concept has its roots in work of why firms need to pursue R&D . . . in order to be able to utilize science-based knowledge.' The notion of absorptive capacity (Cohen and Levinthal 1990) is crucial in R&D activity: networks of innovations could not exist if knowledge cannot be transferred from one entity to another. Indeed, if interaction between the three dimensions leads to a successful set of innovations, the capacity of each firm to transfer knowledge becomes important.

We consider in this chapter the defence industry as a particular case of a sectoral system. As pointed out above, sectoral systems have a product-based view. The defence industry however cannot be fully bounded by a set of products that is common across all defence firms. In order to analyse the specificity of defence firms, it is useful to highlight the links between the three dimensions expounded earlier.

This chapter focuses on the link between the first dimension and the last. Institutional space shapes firms' behaviour. We consider the institutional context as a constraint that is critical in the definition of the overall strategy: qualitative norms for example, or environmental considerations, have strong influences on the behaviour of firms. Thus, the goal of this chapter is to test the possible link between two dimensions defined by the analytical framework provided by Malerba (2002): what is the nature of the relation between the institutional framework and the cognitive one?

The chapter is structured as follows: in section 2, we set out the arguments favouring a 'defence' criterion. The defence industry has some characteristics that bear on the behaviour of firms, and hence on the management and the drivers of innovation strategies and policies. Section 3 presents the dataset used in the study and the theoretical expectations. Section 4 focuses on an econometric assessment concerning the primary hypothesis. Section 5 deepens the analysis of the knowledge base of defence firms. We give the main conclusion in section 6.

2. THE DEFENCE INDUSTRY AS A SPECIFIC STRUCTURE

It is hard to believe that economic considerations are totally independent of the framework defined by the institutional context. As stressed in Johnson (1992): 'Institutions understood as norms, habits and rules are deeply ingrained in society and they play a major role in determining how people relate to each other and how they learn and use their knowledge. . . . In an economy characterized by on-going innovation and fundamental uncertainty, the institutional setting will have a major impact upon how economic agents behave and as well upon the conduct and performance of the system as a whole' (Lundvall et al. 2002, p. 220). Understanding the role of defence institutions in the national innovation system (NIS) is not an easy task but it appears that, even if it is hard to specify the nature of the link between military R&D and economic growth (Guichard 2004), defence institutions are important in many NIS.

2.1 Identifying the Defence Industry

Before explaining the influence of the institutional context on economic life, we must first of all identify the industry itself. Defining sector boundaries is not straightforward because we need to use criteria allowing us to identify all of the actors of an industry.

The first procedure would be to observe the industry and its products. For example, the car industry is just composed of the firms whose production is focused on the production of motor vehicles. But can we add to this category a firm who produces motorcycles? Or tractors?

One of the most important features of industrial organization is that mono-product firms are not frequent. The identification of a mono-product sector (i.e. one composed of mono-product firms) is an easy task since the output is by definition unique and produced by all the firms within the sector. Multi-product firms are more challenging to identify: in this more typical case the product-based definition of industry is not satisfactory per se: we must identify the main activity (or the main cognitive activity); in other words, on which activity the firm's knowledge is especially focused.[1]

Considering this definition, the defence industry is composed of firms whose main activity is dedicated to defence purposes. This definition is limited because we cannot define an exclusive 'defence-dedicated' set of products.

Thus, the defence industry can alternatively be identified by the relationship between firms and governments. Governments are the only clients of defence-related industries, even if firms have a broader activity including civilian markets. In that sense, the notion of dual technology is helpful in the definition of the boundaries of a defence industry.

2.2 Dual Technology

As Kulve and Smit (1993) pointed out, dualism is a hard notion to envelop. The authors emphasize the fact that:

> Historically, the notion entered the discourse on weapons and technology exports that started soon after World War II. . . . Dual use was viewed as a negative feature that complicated export controls: countries might try to obtain military sensitive technologies under the guise of buying civilian technology. The presumed dual nature of some products and technologies also created tensions between the economic and the defence perspective on technology exports, not only within the US, but also between the US and West European countries. (Kulve and Smit 1993, p. 955)

The notion of dualism then switched to take on another meaning after the end of the Cold War. Dual-use technology was seen as a positive feature that should be promoted. The idea was that it could solve two problems: the first related to the difficulty of maintaining a high-tech defence technology base restrained by a limited budget; the second concerning the improvement of a country's economic competitiveness by a more efficient allocation of R&D funds. Governments integrated the notion of dualism into the definition of their policy. The goal of the dual policy is to promote connections between the military and civilian spheres. The problem is that the efficiency of such a policy depends on the conception of dual technology. Two difficulties emerged from this concept: what is technology, and how can we say that a technology is dual.

An interesting view of dual-use technology can be found in Molas-Gallart (1997). His definition is based on the distinction between two dimensions associated with technology. The first considers technology as products and artefacts. In that sense, 'dual-use technology-transfer mechanisms will by definition, refer to ways by which items used in one area of activity can be adapted and used in another' (ibid., p. 369). The second sees technology as including a social dimension. This definition puts the emphasis on the mode of production in which the development and production of artefacts occur. The former conception refers to products per se, the latter concerns the way the products are obtained.

Cowan and Foray (1995) consider that, 'dual technologies are developed and used both by the military or space sectors on the one hand and by the civilian sector on the other'. They also insist on the importance of the social networks in which technology is developed. In this view, dual technology supposes a mixed effort of R&D between civilian and military spheres. Thus, the adoption by defence firms of civilian-developed technology is not a dual technology. The simple use of a technology in either civilian or military domain is not evidence of the existence of dual technology.[2]

For our own theoretical position, the framework provided by Metcalfe and Saviotti (1984) is useful. They consider the case of technological products, with each product defined by three vectors:

1. Vector $\mathbf{Y_i}$ is the performance of a technological product, and measures the benefit associated with the use of technology **i**.
2. $\mathbf{X_i}$ is the technology required to obtain a performance equal to $\mathbf{Y_i}$.
3. $\mathbf{Z_i}$ represents the process to obtain a technological level of $\mathbf{Y_i}$.

Guichard (2004) explains the actions of the dual policy using this triplet. We consider two technologies: *m* is a military technology while *c* is a civilian technology (see Figure 7.1).

The definition of Molas-Gallart (1997) concerns the use of a technology in two or more domains. Thus, we can say that there is a convergence in the case where the same technical levels can lead to a certain performance (say a use) in both the military and civilian markets.[3] In the sense of Cowan and Foray (1995), a dual technology is used and developed by both the civilian and military markets. Thus, there is an integration of the *Z* vectors. Of course, these examples are rather simple and are not supposed to reflect the theories of dual technology expounded by these authors.

The hypothesis in this chapter is that vector *Z* has many determinants. One of these determinants can be the institutional context. Thus, we argue that the ratio of defence to total income is a proxy for the influence of this particular context. The process described by vector *Z* depends on determinants such as social dimension, quality norms, industrial culture, etc. Patents are indeed an indicator of the technical level (*Y* vector). We can observe the influence of the institutional context on *Z* by the ratio

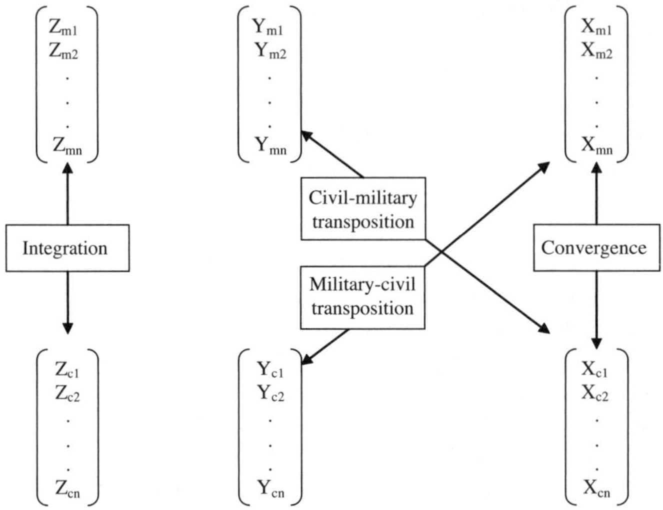

Source: Guichard 2004, p. 112.

Figure 7.1 Technological products and duality

defence/total income and then see if the technical level (*Y*) is shaped strongly or weakly by this determinant.

3. DATA AND THEORETICAL EXPECTATIONS

Following the definition of the theoretical framework, we will present in this section the empirical results concerning the link between defence institutions and knowledge structure.

3.1 Description of the Dataset

We use the dataset provided by *Defence News*[4]. This dataset includes the top 100 defence firms (i.e. firms whose income originates from defence activities) from 1999 to 2005.

Thus, only firms whose income is associated with defence are present. Considering the analysis presented in the previous section, the defence industry is identified as regards the customer. The overall ranking is based exclusively on the defence income.

We have information about the total income, defence income, share of defence income compared to the total, country of the firm[5] and the ranking, which depends on the defence income *only*, for each company. It is important to highlight the fact that not all the firms appear each year, so we may describe these figures as 'quasi-panel' data.

Using the company name, we collected the patents of all the firms using the patent database provided by the European Patent Office (EPO). The EPO includes all the patents applied for in Europe since 1978. Each patent includes useful information: date of application, applicant, countries where the patent is valid, date of publication, title and abstract in some cases. One important feature of this database is the technological class.

Each patent is reviewed by an independent expert from the EPO whose task is to identify the novelty of the invention, and associate the patent with the most relevant technological class. For example, an invention concerning a carrying system for luggage on aeroplanes will be classified in the corresponding category: the classification is then very specific in the sense that careful attention is paid to the use of the final product or the process. Nevertheless, classification is only based on the final use of the innovation, in other words, the purpose of the innovation. Every innovation that concerns, for example, the landing gear of an aeroplane will be classified in the aeroplane category exclusively. Thus, we cannot establish what knowledge is necessary in order to produce landing gear: we only know that the firm has the capacity to produce landing gear.

It is difficult to identify exactly all the items of knowledge required in the production of a product. For example, the production of this type of landing gear may require a physicist to study its aerodynamic features: this knowledge is crucial in the development of the product but cannot be captured by the patent applied for in this technology. Thus, technological classes are a good but imperfect measure of technological knowledge: the knowledge associated with technological classes does not reflect the knowledge dimension within the firm: as Griliches (1990) pointed out, patents are the output of the knowledge production function but are a good proxy of knowledge base of firms.

3.2 Institutional Influence on Patenting Behaviour

Regarding the literature on national systems (Lundvall 1992), sectoral systems (Malerba 2002) or technological systems (Carlsson and Jacobsson 1994) of innovation, we make the assumption that there exists a link between institutional configuration and industrial organization.[6]

What makes sense in our analysis is that each industry has a particular institutional framework that defines its ability to innovate. Thus, the defence industry is strongly linked with an institutional configuration. Defence institutions (Department of Defense in the USA, or the French Procurement Agency) were major actors as they conducted general technological policy. The technological directions were adopted in order to satisfy the constraints of national defence or security. Particular technological fields were developed following a specific demand from defence institutions. As Hartley (2001, p. 90) pointed out:

> Aircraft revolutionized warfare by adding a new dimension to military conflict, leading to the creation of land-based and naval air forces. A new industry also emerged, starting as the aircraft industry, with associated developments in radar and electronics industries. There followed the jet engine, missiles, rockets and space exploration. Some of this new military technology 'spilled-over' into the civilian economy (examples include jet airliners and radar).

4. DATA ANALYSIS

Since the institutional context may influence positively or negatively, we consider that patenting behaviour can differ in the light of the civilian/military trade-off. To assess this major difference, we conduct an econometric test on the data set provided by *Defence News*.

4.1 Prior Analysis

Figure 7.2 presents the number of patents applied for by firms in each year. It is interesting to see that that the number of patents applied for fluctuates over time. There is a steady decrease which we can explain by a shortcoming in the patent database.[7]

Figure 7.3 presents the evolution of average defence income relative to global income (the dotted line is the number of patents). Since all the firms

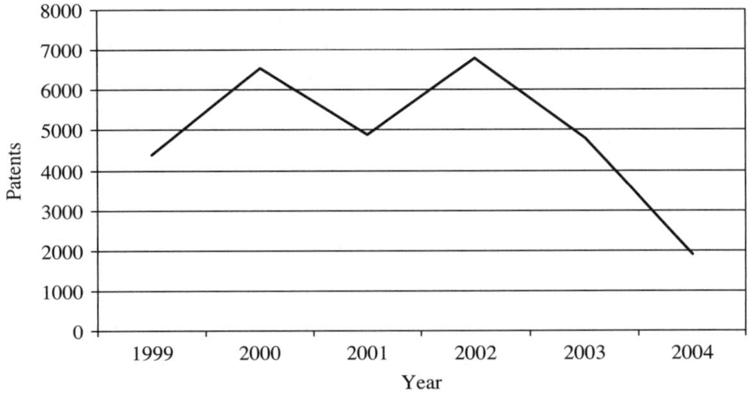

Figure 7.2 Number of patents per year, 1999–2004

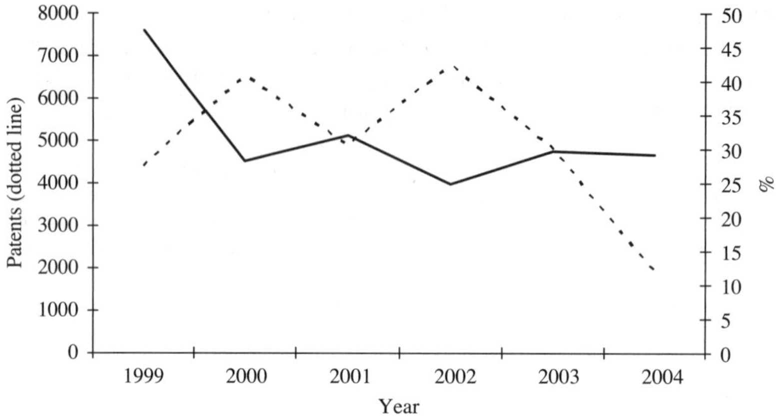

Figure 7.3 *Patents* vs. *share of defence income*

in *Defence News* are defence-related, we can see if there is an evolution of the degree of interrelationship between civilian and military activities. In fact there appears to be a negative short-term relationship between the share of defence in all activity and the number of patents applied for.

Thus, it seems that there is a narrow link between the institutional framework and patenting behaviour. However, the link must be interpreted with extreme caution. We are assuming that the number of patents is a proxy for the knowledge production function. We do not discuss the efficiency of the production of knowledge. Patents are a proxy for technological knowledge in this case.

As pointed out in section 3.1, the database includes the country to which the firm belongs. We present in Table 7.1, a summary concerning the origin of the companies. The table is sorted in ascending order by frequency. We can observe that American companies are over-estimated in the sample but there is no significant influence on the overall results.[8]

4.2 Econometric Tests

The hypothesis tested concerns the potential correlation between patenting behaviour and the share of defence activity (captured through defence income). The following equation is tested:

$$PATENT_{i,t} = \alpha PATENT_{i,t-1} + \beta NOPATENT_{i,t} + \gamma P90_{i,t} + \delta P40M60_{i,t} + \lambda M10_{i,t} + \varphi LNREV_{i,t} + \mu SWEDEN + u_i + \varepsilon_{it}$$

Table 7.1 Frequencies of countries

Country	Frequency	%	Cumulative
China	6	0.54	0.54
Greece	6	0.54	1.08
Indonesia	6	0.54	1.62
Netherlands	6	0.54	2.16
Norway	6	0.54	2.7
South Africa	6	0.54	3.24
Singapore	6	0.54	3.78
South Korea	6	0.54	4.32
Taiwan	6	0.54	4.86
Switzerland	11	0.99	5.85
Australia	12	1.08	6.93
Italy	12	1.08	8.01
Canada	18	1.62	9.63
India	18	1.62	11.25
Sweden	18	1.62	12.87
Turkey	18	1.62	14.49
Spain	24	2.14	16.63
Germany	30	2.69	19.32
Israel	42	3.76	23.08
France	66	5.91	28.99
UK	96	8.62	37.61
Japan	101	9.07	46.68
Russia	126	11.31	57.99
US	468	42.01	100.00
Total	1114	100	

where: $PATENT_{i,t}$ is the number of patents applied for by firm *i* in year *t*;
$PATENT_{i,t-1}$ is the number of patents applied by firms *i* in year $t-1$;
$NOPATENT_{i,t}$ is a dummy that equals 1 if the firm *i* has not applied for any patent in year *t*;
$P90_{i,t}$ is a dummy that equals 1 if the share of defence income is at least 90% for firm *i* in year *t*;
$P40M60_{i,t}$ is a dummy that equals 1 if the share of defence income is at least between 40% and 60% for firm *i* in year *t*;
$M10_{i,t}$ is a dummy that equals 1 if the share of defence income is less than 10% for firm *i* in year *t*;
$LNREV_{i,t}$ is the log of the global income of firm *i* in year *t*;
SWEDEN is a dummy that equals 1 if the company is Swedish;
u_i is the error term for firm *i*;
ε_{it} is the error term for firm *i* in year *t*.

We run the tests with the three 'share' dummies presented here on the ratio of defence income to global income. Regarding the hypothesis concerning the influence of the institutional context on technological strategy, the firm's heterogeneity is observed in the trade-off between civilian and military markets. Patenting should differ between firm A whose defence share may be about 80 per cent and firm B whose defence share approximates 20 per cent. Country dummies are also included in the analysis. There are 24 countries represented in the dataset. The less significant dummies are excluded for the sake of space. Nevertheless, we show the results concerning Sweden, which is the only country that has a strong influence on the outcome: the explanation is that Ericsson has a high patenting policy in our sample (more than 1000 per year) so it tends to bias the influence of Sweden even if the country is not over-represented.

Regressions have been estimated by panel data models (random-effects, fixed-effects and generalized least squares) using different features. We also ran tests using Poisson regression and negative binomial regression regarding the count nature of the explained variable (Nesta and Saviotti 2005). Taking firms from *Defence News*, we then constitute a panel data. There are a total of 485 observations and 165 firms in the dataset. Table 7.2 presents the overall results of the econometric models.

Considering the result of the Breusch–Pagan test for random effects, we can see that the null hypothesis of no random effect cannot be rejected, thus random effects do not provide any significant result concerning individual effects. However, several tests provide information on the nature of the panel data. First, there is a group-wise heteroscedasticity that should be taken into account in any regression. Second, there is a first-order autoregressive disturbance. For these two reasons, we used generalized least square regressions.[9] It is important to stress that our goal is not to estimate all the determinants of a firm's patenting behaviour. Rather, we want to understand the effect, if any, of the civilian/defence dilemma on it. Thus, coefficients should be taken with extreme caution, as well as the estimators. Count data models are also presented to take into account the count nature of our data. As a matter of fact, our discussion will be focused on the signs of the coefficients showing the positive or negative influence on the patenting behaviour.

The *NOPATENT* variable is useful in the sense that there are many firms that did not apply for a patent in some years.[10] This dummy variable allows to control for a potential bias from these observations. One important fact is that we make no distinction between two cases:

1. A firm may not apply for any patent during a year but can be present in the EPO dataset, meaning that the firm 'exists' for the EPO database but has not applied for any patent in that particular year.

Table 7.2 Econometric results

	Random effects	Fixed effects	GLS with heteroscedastic panels and no autocorrelation	GLS with heteroscedastic panels and panel-specific AR(1)	GLS with heteroscedastic panels and common AR(1) coefficient for all panels	GLS with heteroscedastic panels and common AR(1) coefficient for all panels, no intercept
LAGPATENT	0.508	0.001	0.371	0.166	0.194	0.216
	(12.16)***	(0.03)	(8.90)***	(8.54)***	(4.34)***	(6.80)***
NOPATENT	−25.502	−34.872	−9.195	−24.352	−12.593	−28.341
	(−1.56)	(−1.86)**	(−6.36)***	(−15.52)***	(−6.39)***	(−13.04)***
LNREV	15.178	18.226	6.885	15.835	6.597	3.769
	(2.65)***	(0.74)	(7.83)***	(16)***	(4.44)***	(14.15)***
P90	8.279	−4.024	4.694	1.873	1.336	−11.199
	(0.45)	(−0.12)	(3.90)***	(1.25)	(0.51)	(−4.76)***
M10	93.468	7.082	35.775	99.56	25.473	112.491
	(3.58)***	(0.10)	(1.51)	(3.51)***	(0.88)	(3.55)***
P40M60	6.318	3.215	3.241	5.364	0.371	−5.234
	(0.32)	(0.11)	(2.03)**	(3.29)***	(0.16)	(−9.49)***
SWEDEN	273.658		91.188	858.617	146.811	641.907
	(5.67)***		(0.81)	(2.24)**	(0.94)	(1.95)*
INTERCEPT	−102.76	−74.652	−45.775	−101.114	−38.45	
	(−2.12)**	(−0.38)	(−7.88)***	(−13.63)***	(−3.51)***	
R^2 within	0.0002	0.015				
R^2 between	0.7849	0.247				
R^2 overall	0.5896	0.1598				

Table 7.2 (continued)

	Random effects	Fixed effects	GLS with heteroscedastic panels and no autocorrelation	GLS with heteroscedastic panels and panel-specific AR(1)	GLS with heteroscedastic panels and common AR(1) coefficient for all panels	GLS with heteroscedastic panels and common AR(1) coefficient for all panels, no intercept
F test		0.68				
Wald χ^2	556***		196.66***	611.24***	133.15***	452.49***
Breusch-Pagan LM test	0.01					
Log Likelihood			−1792.577	−1606.636	−1675.922	−1628.345
Corr(U_i Fitted)	0 (assumed)	0.2522				
AR Coefficient if available					0.3466	

Notes: All regressions include country dumnies.
Z-stat in parentheses
***, **, * signifigcant at 1, 5 and 10

2. A firm may not 'exist' in the database, meaning that the firm has applied for no patent during the existence of the dataset, i.e. it has a 'zero patent' portfolio.[11]

The *LAGPATENT* variable is significant for almost all the models. The significance of this coefficient suggests that there is an asymptotic convergent evolution *ceteris paribus* of the number of patents. This is relevant to other innovation studies that highlight the fact that there is some kind of inertia in knowledge production activity. The knowledge base evolves smoothly over time: it is difficult to consider that the knowledge base of a firm can evolve in a more erratic way. This hypothesis relies on the well-known notion of path dependency: what firms know today influences what they will do tomorrow. Table 7.3 presents the correlation matrix.

We can also note that the *LNREV* variable, which measures the influence of global income on patenting behaviour, is positive. This 'Schumpeterian' relationship between size and innovative performance is confirmed in our database. As summarised by Nesta and Saviotti (2005, cited by Piscitello 2005), 'several justifications have been advanced for a positive effect of size on the innovative performance of the firm. First, size allows firms to secure the financing of risky activities such as R&D because of the stability of internally generated funds. Second, large firms may spread the fixed costs of innovation over a wider spectrum of sales. Third, economies of scale may arise as several innovation projects are undertaken, thereby increasing their probability of success and reducing the risk involved in one innovation project implemented independently.' Two propositions concerning the defence industry can be implemented following these results:

Proposition 1: Patenting behaviour demonstrates inertia. Considering the t-value of the lagged patent indicator and its statistical significance, we find the number of patents applied for in year t is dependent on the patents applied for in year $t-1$. It could be possible to test for a second-order autoregressive process.

Proposition 2: Patenting behaviour depends on the institutional context. As a matter of fact, institutional context matters. P90 has a negative impact while M10 has a positive impact on patenting behaviour. This confirms the relations presented in Figures 7.2 and 7.3. Defence-related firms tend to patent less than civilian-related firms. This proposition is purely quantitative. There is a correlation between the share of defence income and the willingness to patent.

These two propositions cannot explain the knowledge structure. It appears that the use of patents differs depending on the defence criterion.

Table 7.3 *Correlation matrix*

	LAGPATENT	NOPATENT	LNREV	P90	M10	P40M60	SWEDEN
LAGPATENT	1						
NOPATENT	0.0587	1					
LNREV	−0.3711	−0.688	1				
P90	−0.0321	0.2215	−0.4059	1			
M10	−0.1314	0.0377	−0.0139	0.0352	1		
P40M60	0.2789	0.2429	−0.4468	0.1662	−0.0115	1	
SWEDEN	−0.0541	−0.0187	0.0292	−0.0009	−0.0551	−0.018	1

However, it would be interesting to see the structure of the knowledge base of defence-related and civilian-related firms. This can be done by using the International Patent Classification (IPC) provided by the EPO.

5. KNOWLEDGE BASE

In this section, we will analyse the knowledge structure of some firms. The idea is to observe the cognitive structure of firms depending on the institutional context.

5.1 Technological Spread

Considering the previous econometric test, defence firms tend to patent more than civilian firms. But what knowledge is patented by defence firms? Is knowledge more dispersed or more concentrated when the firm is defence-related? What can we say about knowledge variety in the defence industry (Patel and Pavitt 1997)?

The methodology used is the following. Taking the patents from firms having more than 90 per cent income coming from defence activity, we can see if knowledge is structured in a different way. We believe that concentration will tend to be more important in defence-related firms than in civilian firms.

Table 7.4 presents the firms whose defence/total income ratio is more than 90 per cent[12] for all the years of the dataset. Of the 120 firms that satisfy the P90 criterion, only 43 firms applied for patents during the six years considered by the study.

The methodology presented here will focus on the technological classes. We presume that technological width is less for defence firms since their knowledge is strongly related to defence activities. These activities require

Table 7.4 Technological description using patent information

	Firms	Patents	Technological classes	Herfindahl
1999	3	66	33	551.48
2000	6	80	24	895.1
2001	6	86	25	759.87
2002	9	201	62	423.5
2003	9	257	65	519.47
2004	10	127	49	423.26
Mean	7.17	136.17	43	595.45

a cognitive focus on fewer activities due to the constraints associated with defence institutions.

In order to measure the concentration of knowledge, we use a Herfindahl index. The formula is the following:

$$HT_t = \sum_{i=1}^{n} (s_{i,t}^2)$$

where: H_t is the Herfindahl index for year t
$s_{i,t}$ is the share of technology i in year t

The share in this definition corresponds to the number of patents applied for in technology j compared to the number of patents applied for in all technological classes represented in year t. This 'technological Herfindahl' is oriented to the technological classes as a whole. We do not measure the concentration of knowledge for a firm but want to see if the spectrum of knowledge of defence firms (as an aggregated system of knowledge) is dispersed or concentrated.

It is important to highlight that the Herfindahl index has no constant lower boundary. The upper boundary is always 10000, i.e. the case of only one technology applied for in year t. The lower boundary of the index depends on the number of technologies. If there are 20 technologies, the lower boundary will be 500, etc. Thus, the index is a decreasing function of the number of technological classes.[13] Figure 7.4 shows the evolution of the technological Herfindahl index for defence-related firms (dark line) and civil-

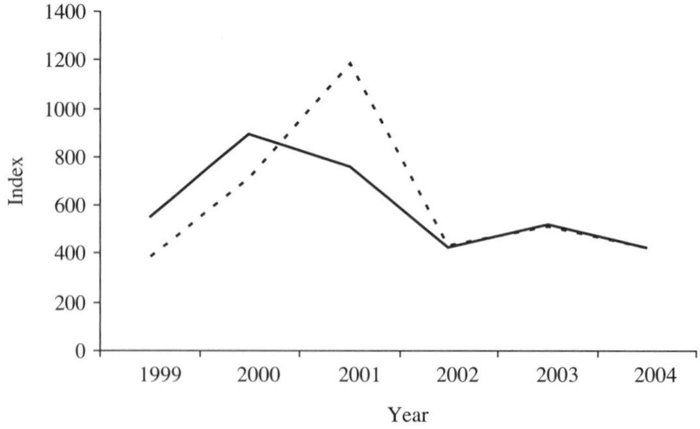

Figure 7.4 Evolution of HT for civil (dotted line) and defence firms, 1999–2004

related firms (dotted line). The HT index is a proxy for the knowledge spread for a particular institution. There is no evidence of any difference between the knowledge base of defence firms and civil firms considered as a whole.

5.2 Knowledge Concentration of Defence Firms

We next assess the use of a Herfindahl index associated with firms. This index will be noted HF. The idea is to capture the degree of concentration (or cognitive focus) of defence-related firms. Each firm has a patent portfolio per year. This portfolio can be wide or narrow depending on many factors such as firm size, R&D expenditures or other. Our argument is that the difference in patenting behaviour is due to a different institutional configuration. To some extent, it is possible to argue that the institutional framework shapes not only patenting behaviour but also what is being patented. We suggest that a defence firm's knowledge is more specific than civilian firms. This is measured by a similar index derived from the following formula:

$$HF_{j,t} = \sum_{i=1}^{n} (s_i^2)$$

where s_i is the share of patents applied in technology i by firm j.

This index captures the focus of knowledge: a firm which patents only in one technology will have an index of 10000. Table 7.5 shows the HF index of defence-related firms per year.

We can see that the mean HF index ranges from 2097 to 5885. The table gives rise to two major observations:

1. Knowledge of defence-related firms seems to be highly concentrated around a few technological classes. This confirms the hypothesis presented in the beginning of the chapter.
2. Large firms tend to have a more dispersed knowledge. The global income of large firms is positively correlated with the number of patent applications and the knowledge spread. This can be explained by the Schumpeterian analysis expounded earlier.

6. CONCLUDING REMARKS AND RESEARCH PERSPECTIVES

This chapter has proposed a test of the link between the institutional context and patenting behaviour. It appears that there is a negative relationship between the share of defence in global activity and the quantity of patents applied for at the EPO. Thus, the notion of a sectoral system of innovation

Table 7.5 Technological concentration index of defence-related firms

	Firm	Herfindahl	Mean	Country	Total Income in millions of US dollars, 1999
1999	BAE Systems	1009.37	2097.77333	UK	11686
	GIAT Industries	1720.23		France	1281
	United Defense	3563.72		USA	1200
2000	General Dynamics	2000	2718.896	USA	8950
	GIAT Industries	1238.3		France	827.5
	Krauss-Maffei	1836.74		Germany	632.9
	Oerlikon	1644.44		Switzerland	348
	United Defense	6875		USA	1213.5
2001	Elbit Systems	5555.55	4079.48	Israel	519.1
	GIAT Industries	1712.81		France	474
	Krauss-Maffei	1961.6		Germany	701.4
	Oerlikon	3580.25		Switzerland	1026.1
	Rafael Armament	1666.67		Israel	675
	United Defense	10000		USA	1200
2002	DRS	10000	5885.27556	USA	517.2
	Engineered Support System	10000		USA	390.5
	GIAT Industries	650.11		France	701.7
	Israel Military Industries	10000		Israel	435.4
	Krauss-Maffei	3933.52		Germany	593.6
	Lockheed Martin	583.85		USA	23990
	Newport News Shipbuilding	10000		USA	2000
	Rafael Armament	5000		Israel	760
	United Defense	2800		USA	1300
2003	Engineered Support System	5000	2984.59444	USA	407.9
	Krauss-Maffei	4591.84		Germany	631.1
	Rafael Armament	2222.22		Israel	760
	GIAT Industries	2243.77		France	814.5
	United Defense	2653.06		USA	1725
	Titan	1020.41		USA	1392
	Rheinmetall	2868.48		Germany	1758.1
	Raytheon	706.01		USA	16760
	Elbit Systems	5555.56		Israel	827.5

Table 7.5 (continued)

	Firm	Herfindahl	Mean	Country	Total Income in millions of US dollars, 1999
2004	DCN	3333.33	4533.898	France	2084.5
	DRS	2500		USA	1001.3
	Elbit Systems	10000		Israel	898
	Engineered Support System	10000		USA	572.7
	GIAT Industries	3333.33		France	915.4
	Krauss-Maffei	2400		Germany	791.1
	Lockheed Martin	603.92		USA	31824
	Rafael Armament	2500		Israel	830
	Raytheon	668.4		USA	18100
	United Defense	10000		USA	2056.2

can be applied to the defence industry. It seems that the innovation system of such an industry is driven by specific goals that are defined by the institutions: the technological path seems more targeted for defence-related firms.

We have not found any real difference between the technological spectrum of defence-related and civilian-related firms. The next step has been to construct the firm Herfindahl index for civil-related firms and check their knowledge concentration.

Another research path could be the integration of the notion of corporate coherence. This notion measures the evolution of the knowledge base of a firm considering the technological path that is taken. Firms can apply for patents in multiple technologies: the coherence measures the relationship between these technological fields. It could be possible to argue that corporate coherence is different in the case of civilian firms than military firms. The construction of a more relevant concentration index should also lead to better understanding of the phenomenon. In fact, the HF index can equal 10000 in the case of a firm that has applied for one patent in the period. The new index should include this bias in order to distinguish between massive patenting behaviour and concentration of technological classes.

NOTES

1. The core competencies view is in fact a possible answer to this problem (Amesse and Cohendet 2001, p. 1469).

2. The purpose of the authors was focused on the understanding of technological transfer (spill-over) between civilian and defence firms.
3. A computer for example can be developed by a civilian firm and later used in military firms. There is a cost related to the transfer of a technology from civilian to military domains. The quality norms associated with military activities are different from civilian markets: there is a cost of 'militarization'.
4. Available on the *Defense News* website www.defencenews.com.
5. The country reflects the residence for tax purposes.
6. For a long-term perspective of the co-evolution between institutions and economies, see Murmann (2003).
7. The database goes from January to April 2004.
8. Country dummies are included in the econometric test and show no significance for the USA.
9. The results of these tests can be obtained on request to the author.
10. In fact, this dummy is not used in the count models because they take into account the particular shape of the distribution.
11. Of course, it is possible to say that this could be an error of the dataset or that the firm has a more localized patenting behaviour.
12. Firms that satisfy the condition implemented by the P90 dummy.
13. This Herfindahl index is based on technological classes across firms. The hypothesis relies on an aggregated system of knowledge which is common to all the firms. We will also consider the Herfindahl index for firms (concentration of technological classes for a specific actor).

REFERENCES

Amesse, F and P. Cohendet (2001), 'Technology transfer revisited from the perspective of the knowledge-based economy', *Research Policy*, **30** (9), 1459–78.

Carlsson, B. and S. Jacobsson (1994), 'Technological systems and economic policy: the diffusion of factory automation in Sweden' *Research Policy*, **23** (3), 235–48.

Cohen, W.M. and D.A. Levinthal (1990), 'Absorptive capacity – a new perspective on learning and innovation', *Administrative Science Quarterly*, **35** (1), 128–52.

Cowan, R. and D. Foray (1995), 'Quandaries in the economics of dual technologies and spillovers from military to civilian research and development', *Research Policy*, **24** (6), 851–8.

Graham, S. (2004), '"Hiding in the patent's shadow": firms' use of secrecy to capture value from new discoveries', GaTech TI:GER working paper series, accessed 12 April 2007 at http://tiger.gatech.edu/files/gt_tiger_hiding.pdf.

Griliches, Z. (1990), 'Patent statistics as economic indicators: a survey', *Journal of Economic Literature*, **8**, 1661–707.

Guichard, R. (2004), *Recherche militaire: vers un nouveau modèle de gestion?*, Paris: Economica.

Hartley, K. (2001), 'The RMA: an economist's view', in R. Matthews and J. Treddenick (eds), *Managing the Revolution in Military Affairs*, New York: Palgrave.

Jensen, M.B., B. Johnson, E. Lorenz, and B.-Å. Lundvall (2004), 'Absorptive capacity, forms of knowledge and economic development', paper presented at the Second Globelics Conference in Beijing, 16 October.

Johnson, B. (1992), 'Institutional learning', in B.-Å. Lundvall (ed.), *National Systems of Innovation: Towards a Theory of Innovation and Interactive Learning*, London: Pinter Publishers.

Kulve, H.T. and W.A. Smit (2003), 'Civilian–military co-operation strategies in developing new technologies', *Research Policy*, **32** (6), 955–70.

Lundvall, B.-Å. (1992), *National Systems of Innovation: Towards a Theory of Innovation and Interactive Learning*, London: Pinter Publishers.

Lundvall, B-Å., B. Johnson, B. Dalum and E. Andersen (2002), 'National systems of production, innovation and competence building', *Research Policy*, **31** (2), 213–31.

Malerba, F. (2002), 'Sectoral systems of innovation and production', *Research Policy*, **31** (2), 247–64.

Metcalfe, S. and P.P. Saviotti (1984), 'A theoretical approach to the construction of technology output indicators', *Research Policy*, **13** (3), 41–51.

Molas-Gallart, J. (1997), 'Which way to go? Defence technology and the diversity of dual-use technology transfer', *Research Policy*, **26**, 367–85.

Murmann, J.P. (2003), *Knowledge and Competitive Advantage*, Cambridge: Cambridge University Press.

Nelson, R. and S. Winter (1982), *An Evolutionary Theory of Economic Change*, Cambridge, MA: Harvard University Press.

Nesta, L. and P.P. Saviotti (2005), 'Coherence of the knowledge base and firm innovative performance: evidence from the US pharmaceutical industry', *SPRU electronic working paper series*, no. 113, University of Sussex.

Patel, P. and K. Pavitt (1997), 'The technological competencies of the world's largest firms: complex and path-dependent, but not much variety', *Research Policy*, **26**, 141–56.

Piscitello, L. (2005), 'Corporate diversification, coherence and firm innovative performance', *Revue d'Economie Industrielle*, **110**, 127–48.

PART III

From knowledge to market: systemic links at the local level

8. The logic of the innovative milieu

Dimitri Uzunidis

Technological innovation is the safest way to restore, transform and enlarge markets. The extension of companies as well as the increasing globalization of markets reveal the importance of local pools of productive resources. The geographical proximity between science, technology, industry and finance contributes to the emergence of innovations. Interactions are being organized through the interplay between private actors and political institutions. Today, economists consider the 'local economy' as a geographical and economic platform for the organization of production and, as a consequence, as an opportunity to create new activities, goods and services, new employment and sources of income. For almost four decades, the innovative approach based on proximity and, in particular, on the concept of the 'innovative milieu' has demonstrated its pertinence as a form of modelling of decentralized economic growth, but also as a source of enrichment for firms' technological competencies. These economists are not part of a strictly linear approach to innovation (that would correspond to the view that an increase in inputs – in this case R&D expenditures – leads to an increase in outputs – namely innovations), although they identify the need to increase R&D expenditures in order to enlarge the knowledge base. They are rather part of an interactive vision that stresses the importance of networks of private and public actors at a specific level – the region – considered as pertinent for the implementation of innovation policies in the specialized literature dealing with economic geography. This literature includes the works of Aydalot, Becattini, Courlet and Pecqueur, Rallet, Gilly, Torre and Dupuy, Veltz, Benko and Lipietz (for a synthesis, see Hèraud 2003).

In point of fact, in a constantly changing economic environment that is extremely difficult to foresee, the company, small or big, constantly arouses the attention of politicians, economists and sociologists. Its main function – innovation – is viewed as a predominant source of employment, wealth and prosperity. Two conditions must be fulfilled to play this role: (a) its structures must be flexible enough so that the enterprise will be able to adjust itself to the hazards of the market; (b) the constant renewal of its productive resources will only be achieved if the financial and industrial capacity of a country or region (and even more, of a local economy) is sufficiently strong and diversified to enable an enterprise to include producers' and consumers' networks in the constitution of its supply as well as in building sufficient demand for its products. When these prerequisites are respected, the creation of small innovative enterprises and the strengthening of the innovation potential of big companies may be at the core of industrial policies and planning. Big enterprises group various production, finance and marketing activities and seek to organize market opportunities and technological development through alliances, mergers and buyouts. This strategy is enriched by the absorption of – or the cooperation with – small enterprises which complete innovation systems.

An innovation system describes the relationships (scientific, technological, industrial, commercial, financial, political) between private and public institutions (enterprises, research and engineering labs, administrations . . .) (see works by Freeman, Lundvall, etc.). In general, the relationships consist of financial and information flows and the movement of persons. The purpose of that system is to produce innovations (new organizations, new goods and processes, new resources, new combinations of productive resources). Systems are national or local; in this case, the stress is put on public legal regulatory mechanisms. Systems may also be 'private'; in this case the analysis gives the preference to a 'network' which may be defined as a set of enterprises legally and/or financially linked to one or several bigger companies (parent company). A network is a system whose purpose is to achieve integrated production (often multiple) in a single value chain and under the direction or coordination of holding or central firms (Uzunidis 2003).

An analysis starting from the innovative milieu makes it possible to study the firm's environment in order to understand its innovation dynamics. In fact, the systemic nature of the relationships that characterize an economic and social milieu makes it possible to identify what contributes (or not) to the innovative act. However, do we need to reduce innovation (as a product of the milieu) to merely inter-personal exchanges that lead to a new productive combination? Is it only the result of a specific organization of economic relations? Our argument is that the systemic nature of the milieu does not exclusively relate to economic interactions but more precisely also takes into consideration the social structures that are at the origin of innovative behaviours. Moreover, institutions (states, local communities) play an important role in the organization and development of socio-economic structures. In its turn, the innovative milieu contributes to the company's innovative performance through the supply of scientific and technological resources.

1. PROXIMITY AND THE INNOVATIVE MILIEU

1.1 Economic Proximity

The concept of 'proximity' is now widely used in both industrial economics and innovation studies (see notably Rallet and Torre 2002). However, the variety and scope of its applications ('proximity economies') mean that its use must be adequately studied. A priori, proximity seems to be related to the existence of localized externalities generating phenomena of spatial concentration and regional dynamics. That first meaning of physical proximity was soon supplemented by other interpretations (listed by Sierra 1997) in which the operating field of proximity (space, organization or institution) is intertwined with the contents of the proximity relationship (information, training, knowledge, technology, etc.). Without entering into a critical debate that would lead us too far afield, in order to be consistent we must give priority to the domain in which the proximity relationship takes place (before defining its content). From this perspective two types of proximity are usually considered when the issue of localization is coupled with the organizational problem: spatial-temporal proximity and organizational proximity.

The specific nature of the activities involved in the production of new knowledge and the interactions associated with them have led economists to introduce the concept of cognitive proximity in addition to spatial-temporal proximity and organizational proximity (Nooteboom 2002). Cognitive proximity means the more or less formalized sharing of experiences, codes, languages, models resulting from and facilitating the communication of information inside organizations. Due to its nature, cognitive proximity occupies a special place in research activities and is also behind various communication flows inside and outside the company. In the case of interactions related to industrial research, cognitive proximity impacts not only on the internal relations of research centres, but also on external interactions between those centres and the other departments of the firm and on interactions with the environment (other labs and partners in the field of research and innovation).

Therefore, we propose a three-dimensional approach to proximity before stressing the importance of spatial proximity in the creation of new activities and the launching of innovations. Table 8.1 identifies the main parameters characterizing the different categories of proximity as well as the operating field and the types of stakes related to them. It is worth noting that interactions are generally multi-dimensional: they represent a combination of different dimensions from which a major dimension emerges. In our case, this core dimension relates to space and time.

Table 8.1 The three categories of proximity

Proximity	Parameters	Operating field and stake
Spatial-temporal	Distance vs speed	Displacement : flow
Organizational (intra- and inter-organizational)	Hierarchy vs market Intra-firm vs extra-firm Vertical vs horizontal Instruction vs Contract	Co-ordination Strategy, actions, routines
Cognitive	Code vs content Context vs understanding (awareness and interpretation)	Communication Concept, ideas, knowledge

A local economy (or a local production system) may be defined as a geographical area consisting of a set of systemic relationships among enterprises and also between enterprises, public authorities (the state) and local communities; these systemic relationships characterize the area localized for a given type of activities or final production. That economy is observed and studied as a node of productive relations which contributes to regional and local planning (remaining national). Economists also deploy language in order to show the dynamics of the combinatory and complementary relationships between enterprises of different sizes at the local level, e.g. 'local productive system', 'innovative milieu'.

Alfred Marshall (1919) demonstrated that the regional efficiency of such an organizational mode resides in what we call today economies in transaction costs. The concentration in a single geographical location of the main actors of the same productive system (mainly producers on one side and users on the other) not only facilitates transactions, but the mutual relations of knowledge and confidence between different partners. The development and accumulation of expertise will therefore create what Marshall called the 'industrial atmosphere' facilitating the functioning of the local labour market. This phenomenon is related both to the workers' qualifications and experience and to the location of several enterprises in the same locality. In the meaning attributed by Marshall to the 'industrial district', this environment includes a specific density of population with proven qualifications, a set of actors constituting the different links of a single production system, and finally a degree of know-how strongly resulting from acquired experience. The main characteristic is that the local organization of production is not linked to a hierarchical principle regulating an enterprise, but is rather based on a principle of collaboration and cooperation between different production units. Therefore, the concept of solidarity between economic actors is of considerable importance. The local production system is mainly characterized by the proximity between productive units (individual firms, service suppliers, research centres and training institutions and so on). The relations between these units have a variable intensity and may take on highly differentiated forms: formal or informal relations, market or non-market, and so on. Alliances mainly relate to the flow of materials, services, labour, technologies and knowledge.

Initially, it was the GREMI (Groupe de Recherches Européen sur les Milieux Innovateurs/European Research Group on Innovative Milieux, a research team belonging to the Economic & Social Research Institute of the University of Neuchâtel, Switzerland) which started, in 1985, from the assumption that regional milieux are those that primarily diffuse the different types of innovation (product innovations, process innovations, organizational innovations, social innovations, training/qualification innovations, etc.). The emergence of a 'successful' region results from the fact that it is able to manage its own capacity to develop new products, new techniques and new organizations. As one of the founding elements of regional science, this assumption questions traditional economic theories which, in the opposite way, were favourable to forms of progress and growth whose triggering factors are to be mainly found in and justified by 'external' influences (so-called models of 'top-down development').

Our central theoretical assumption concerning the concept of the innovative milieu, namely the social and economic environment of a region developed over the course of history ('path dependence') is that all innovative milieux are the product of interactions between firms, institutions and labour. Such interactions are exclusively the result of mutual synergies (networks, partnerships, and so on) between different local agents (public or private) participating in economic and industrial development. For example, we may refer to the different forms of cooperation between enterprises and research labs. It is firstly the surrounding socio-economic industrial and scientific environment that contributes to the creation of new activities (in particular through entrepreneurship and spin-off) and to the genesis of innovations (see also Konstadakopoulos 2000). In addition, this can only develop in a 'natural' form if some preconditions are respected, among which are: existence at the 'local' level of a community of actors (enterprises, research and training centres, public administrations, professional qualifications, and so on); presence of material, human, financial, technological and information resources in the immediate geographical neighbourhood; existence of specific know-how leading to high-quality productive activity; existence of relational capital favourable to the creation of local, national or international networks; and the existence of norms, rules and values promoting positive behaviour among economic actors.

The concept of the innovative milieu also underlines a powerful principle concerning innovation systems: it reinforces the claim that the innovation capacity of a firm is closely related to the social, economic and political factors surrounding it. In this case, the expression 'innovative milieu' generally relates to the capacity of a local economy to generate innovations through the emergence of new enterprises and the location of already established firms in its geographical area, where the industrial commercialization of research guides the creation of small innovative enterprises.[1] The local economy is therefore represented in the form of a spatial system valuing all kinds of capital and merchant exchange. Endowed with a certain degree of organizational and economic autonomy, the leading characteristic of that local economy is constituting and valuing specific resources and implementing specific combinations of resources. The latter are composed of capital and labour of specific forms and content and applicable to particular activities and sectors: they are specific at the technological, financial or demographic levels, but also at the level of competences, qualifications, levels of education, etc. Thus, the local economy becomes an 'innovative milieu' and reduces the risks related to the uncertainty of a given investment; it triggers an innovation process that includes the creation of enterprises and the incorporation of already existing technological enterprises.

1.2 Company Strategies and the Innovative Milieu

In order to understand and study an enterprise, certain economists look at the internal organization of production units and their environment (market competition, management, and so on). Their systemic approach leads them to consider the firm as a living organism whose birth, growth, survival or death are regulated by a set of contradictory relations between the unit and its environment as well as among the internal organs of the unit. Economists go so far as to assert that a firm as such has no meaning: what matters are its relations with other companies, markets or institutions (Robinson 1972). This image of the company stresses the role played by the commercial, financial or technological relations generated or borne by the enterprise. In this way, we may finally evaluate the role of a local innovation system or 'innovative milieu'.

An enterprise is usually defined as an economic unit, a set of combined production factors whose activity leads to the manufacturing of goods and the supply of services sold in a market. Its objective is the realization of profits essentially and durably superior to those made by its competitors and sufficient to finance its investments and growth. Increasing its size is another condition that should protect the firm against the attacks of competitors and the fluctuations of demand. The reality of the firm is complex: from a legal point of view, an enterprise is an autonomous decision-centre; its implements a strategy, it defines objectives and avails itself of the means necessary to reach its purpose. The enterprise is also a social organization, gathering people with different competencies, and linked by hierarchical relations of power and responsibility. The economic autonomy of the enterprise is rather limited. On the one hand, the constraints imposed by its legal and commercial environment restrict its margin for manoeuvre. The need to make a profit guaranteeing good health leads it to enter into conflictual situations or to cooperate with other enterprises; to change its status whenever its social capital increases; to protect, diversify and increase its markets. On the other hand, its capital sometimes belongs, totally or partially, to another company: subsidiaries and other enterprises subject to sophisticated links or financial integration depend on decisions taken by the group to which they belong.

Decision and power are the two keywords as regards the functioning of an enterprise. The firm's decision-system governs its activities. The system is based on the power and control exerted by the owners of the enterprise and is used to determine the decisional power of its executive director and management. Generally speaking a decision is made at three levels: operational (e.g. production activities in the firm), managerial (organizational tasks, monitoring and procedures), and strategic (planning, programming, global policy). The organization charts corresponding to this scheme reflect the management structure of the enterprise. The theory of organizations (Williamson 1985) teaches us that the organization chart of an enterprise reveals its hierarchies, functions and activities in relation to the place of decision-making, information structure, consistency factors and the dynamic core of functioning. The different functions presented in the organization chart are interrelated by: (a) instruction flows that circulate from top-down between three decision-making levels and express a certain technical and decisional rationale; (b) flows of technical, economic and financial information circulating between departments and subsidiaries; (c) financial flows generated by the budgeting of the various activities of the services, departments and subsidiaries and the contracting activities between the parent company, subsidiaries or with other interdependent firms.

The enterprise may be compelled to increase its size in order to survive. In fact, it must partly determine its future in order to fulfil the expectations of its shareholders (in the best conditions possible as far as profits are concerned while mastering financial and commercial risks). For that purpose, the enterprise must reduce the uncertainty that characterizes the functioning of the market and avail itself of all the necessary means to capture, screen, process and use the greatest amount possible of economic, technological, financial, commercial and political information. A strong change in the environment leads to a quicker turnover of capital, an acceleration of the pace of innovation and increased risk of failure in the process of 'creative destruction'. In this case, information becomes a powerful weapon in the competitive struggle (Laperche 1998). Some 70 years ago, Coase (1937) raised the issue of the real existence of the firm. Transaction costs are so heavy to bear that the enterprise must on the one hand control the situation by stabilizing its productive resources, and on the other hand protect itself behind any barriers to entry of new competitors it has itself erected around its market. Therefore, competition not only becomes imperfect but evolves into a power game (combining competition and temporary agreements relative to alliance and cooperation) between worldwide industrial and financial groups. The room for manoeuvre left to small entrepreneurs is therefore marginal and unsteady.

The analysis, in particular in the field of innovation management, focuses on centralization versus decentralization, by opposing factors that favour one or the other of these alternatives (e.g. economies of scale, cross-fertilization, tacit communication of information, contacts with production and marketing departments, control over technological assets, etc.; or, at the opposite end: access to rare competences, external economies in a growing environment, proximity to clients or suppliers, etc.). In this perspective, the choice of the location of the emergence of new activities is determined by the relative weight of the centralization/decentralization factors. The nature of activities (technological level) strongly determines the level of compromise between, on the one hand, the search for externalities (agglomeration effect) and, on the other hand, constraints that are related to the scale and the indivisibility that implies concentrating innovation resources at the local level.

As a matter of fact, the constant broadening and renewal of markets, resulting from the combined evolution of profit and financial returns, increases commercial risks, as reflected in the number of products withdrawn from the market even before those investments made to produce them have been amortized, and even before they are known by consumers. They also increase financial risks; many firms have gone bankrupt – e.g. Internet firms in the USA and Europe – due to speculation and indebtedness even before they penetrated the market, because of a lack of customers. In such cases, the enterprise is compelled to invest huge amounts of capital to constitute a powerful information system, to be ready to fight against such risks, constantly innovate and reassure its shareholders, financiers and politicians in different countries and attract customers while ensuring their loyalty.

The process of 'permanent innovation' is the main characteristic of 'global firms'.These are defined as enterprises whose integration is determined by multiple information and financial flows and whose structures are widely de-concentrated: network firms with multiple subsidiaries, multiple partnerships between sub- and co-contractors; enterprises with huge financial capacities in an industrial or service sector characterized by a high scientific and technological capacity and having major external impact. The 'global firm' has a strong capacity to innovate and to modify its structures and organization mode constantly. It benefits from the comparative and specific advantages resulting from various sites (locations); it integrates in a value chain all the separate activities it performs worldwide (R&D, logistics, innovation and financial engineering, manufacturing, assembly, marketing, various services).

In addition to the structural and market advantages conferred on them by public authorities and local communities, firms benefit from a de-concentrated management structure enabling the location of the innovative activities appropriate to a worldwide company. The firm itself gives meaning to the development of the region in question (Figure 8.1). Due to the fact that markets are both open and concentrated, but also characterized by differentiated and diversified demand (for high, medium and low incomes), a big company follows a global and permanent innovation policy. This leads it to invest in the continuous enrichment of its innovation capacity. Regions with a strong scientific and technological potential succeed in attracting those enterprises which, through a strategy aiming at the diversification of their portfolio of securities and their relations with partners, consider that the 'cluster' is a privileged way of accessing the new

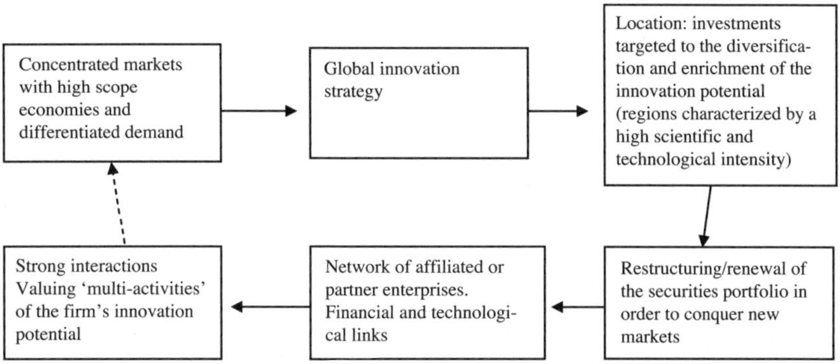

Figure 8.1 Location of scientific and technical activities of the innovative global firm

productive resources accumulated in a given locality. Is it therefore possible to argue that the innovative milieu integrates the accumulation domain of the firm or of several firms belonging to the same cluster?

The regional anchorage of the enterprise enables it to avail itself of a pool of resources (and sometimes a market) in order to amortize the costs inherent in its investments, in an economy undergoing constant change. However, this regional anchorage depends on the quality of the 'pool' mentioned above, compared to the firm's expectations as regards innovation and commercial expansion. Hence arises the necessity for governments and local authorities to create a system, organizing the resources with a view to generating multiple innovation processes taking into account the competition–cooperation behaviours between the same actors, in an open economy (see in particular Amable et al. 1997; Pitelis et al. 2005). This system is a supplier of those productive resources that are capable of generating technological entrepreneurship and attracting big enterprises with confirmed performances in the innovation field (Uzunidis 2004; Madeuf et al. 2005).

2. ACTORS AND ACTIONS

2.1 Policies Related to the Emergence and Attractiveness of Innovating Enterprises

Since the end of the 1980s, the financial, commercial and production strategies of the 'global firms' have replaced managed country planning; at the same time, planning policy has become obsolete. In fact, the opening up of economies to competition (deregulation and contestability of markets aiming at the suppression of monopolistic practices) has led to a strong overlap between national economies, to the point that the objectives of the big international companies have become the national interest. For example, in the last few years and as far as France is concerned, public authorities have intended to rehabilitate the concept of industrial policy, now centred on services and intangibles (Beffa 2005; Gaffard 2005). While the major industrial programmes of the 1960s were essentially domestic and sectoral (the 'machine-tool plan', 'information plan', etc.), recent developments concerning state intervention have acquired an exclusively regional dimension. Instances include the aim of DATAR (the Delegation for National Planning and Regional Action) in 1998 to list and promote the maximum number of local production systems and industrial districts, and more recently the initiative concerning poles of competitiveness. These governmental initiatives, often preceded or followed by similar initiatives adopted by regional authorities (support given to clusters becomes a leitmotiv in all regions) are of specific interest to us because they confirm or clarify trends that were already at work concerning the role of the spatial dimension in the functioning of productive processes.

The role of the state in the formation and organization of scientific and technological supply leading to innovation and investment is both indispensable and justified. Public intervention has already gone beyond the traditional domain of the simple implementation and financing of a scientific and technological policy having at its centre public educational and research institutions. The behaviour of the state in this field intermingles more and more with that of the big industrial and financial groups, and the strong relations of interdependence existing between these influential organizations justify the transfer of resources from the public to the private sector. The instrument used for that purpose is the formulation by the state of an innovation policy, namely promoting all the scientific means of research and development, and encouraging the elaboration of new products and new processes at the industry level.

The intervention of the state and local decision-makers in the creation and management of an 'innovation system' may take various forms: financing activities to generate resources that may be acquired individually or collectively by enterprises; creating mechanisms enabling a re-appropriation by enterprises of investment in R&D (e.g. patents that do not hamper the diffusion of innovations); or implementing cooperative procedures between public and private entities in order to guarantee the financial feasibility of a private investment project capable of having large-scale economic consequences (for clear summaries of theories of 'endogenous growth', see Guellec and Ralle 1995; Aghion and Howitt 1998).

Building up a pool of productive resources that may be tapped at any time by enterprises is presently viewed by economists as the fundamental aspect of the intervention of the state in the economy and organization of the space. Branscomb and Keller (1998) note that the creation and diffusion of knowledge increase the performance of a national or local economy (and of the enterprises constituting it); they suggest that traditional scientific and technological policy (oriented towards financing and implementing huge R&D programmes, essentially in the fields of defence, energy, space or medicine) has been replaced by a policy devoted to research and innovation. In order to be successful from the point of view of competitiveness, this policy must aim both at the implementation of public research programmes or of projects financed by the state and at the diffusion of their results to users (enterprises). The state must guarantee the efficiency of the procedures of research commercialization on the basis of regulations (protection of industrial property, antitrust laws, etc.), fiscal procedures, budgeting, etc., with the purpose of triggering propagation effects. In this way, economists conceptualize a 'stock' (pool) of resources that are made available in the course of that multi-shaped, multifunctional and multi-partner co-operation.

In a network economy, local economies are now seeking to develop by relying on private initiatives coupled with targeted public action. Investment attractiveness, the capacity to establish enterprises and the creation of jobs determine the performance of a local economy. These three indicators are linked both in time and in space. However, regional institutions put forward a certain number of arguments to attract direct investments with the potential to generate employment in the short term. One distinguishes two types of policies concerning the growth and promotion of investments: short-term and structural.

'Short-term policy' brings us back to budgetary and fiscal policies whose purpose is to attend to the most urgent problems: establish enterprises in order to create jobs, i.e. attract production units to generate employment. The results expected in the field of investment must be achieved extremely quickly. Public authorities concentrate their attention on: (a) enterprises with mobile production units (Fordist or heavy, in which total manufacturing costs determine the choice of an investment); (b) enterprises with low value-added that already have a basic trade capable of being immediately capitalized. The main measures of such an economic policy are the following:

Financial incentives Direct allocation of funds to the investor (enterprise and entrepreneur) by the state or local authorities (investment premium and subsidies, employment subsidization, preferential loans);

Tax incentives Reduction of the investor's global taxation (temporary tax and employer's contribution exemptions, relief of customs duties on raw materials, intermediary goods and capital goods);

Indirect incentives Grants to the investor of land, buildings, telecommunications (cf. Enterprise zones), privileged access to public markets, flexible part-time recruitment for a limited period, etc.

The concept of 'structural policy' brings us back to industrial and innovation policies that aim at constituting or maintaining strong economic and technological specialization: enrichment of the scientific and technological capacity in order to trigger the creation of innovative enterprises and to attract research and production centres into specific fields of specialization of the large enterprises. The results are cumulative and long-term; public authorities avail themselves of tools enabling them to organize permanent technological and commercial monitoring in order to guide decisions pertaining to the creation, restructuring and enrichment of innovation networks (investments and marketing).

The main measures of a structural investment policy are the following:

Major public investments for the creation of conditions favourable to long-term endogenous growth Transport and communication infrastructures; effective educational, research and engineering structures; local financial systems oriented towards innovation; effective health systems; quality of life objectives through cultural space planning and leisure activities, etc.

Implementing and financing a research and innovation policy In place of a policy strictly dealing with industry and services – with the following objectives: (a) conjunction around a specific programme based on the competences of enterprises, public and private research institutions, associations and decision-making bodies; (b) networking among the different actors involved in research, industry, trade and forecasting in order to inject added-value investments in the field (and its niches) as defined by controlling authorities.

Creating a public pole supplying services and capital For enterprises attracted by the project and entrepreneurs involved in core and ancillary activities (information engineering, national planning and socio-economic studies), e.g. creating an observatory for the local economy endowed with genuine consultative and organizational power. For example, in the case of global companies, the management of research, development and production aims to define the global strategic orientation of a de-centralized R&D – searching for a scientific environment or sticking to the most dynamic markets – through multiplying and deepening local links with different partners (Cantwell and Janne 2000; Dupuis and Gilly 1996). Thus, globalization is not opposed to the strengthening of local interactions but implies, on the contrary, the search for and mobilization of such interactions (hence the importance of research on proximity economics and spatial innovation).

2.2 Innovative Environments and the Emergence of Innovating Enterprises

At the core of a local innovation system, specific attention can be paid to the creation of small innovating enterprises. In the present economic uncertainty and following the trends to de-centralization, the establishment of enterprises is supposed to solve many problems linked to the rigidities resulting from institutional interventionism. Its flexible structures enable much easier reaction to consumers' expectations. Moreover, the possible bankruptcy of an enterprise does not threaten the financial and industrial fabric of the country, region or place. Establishing a considerable number of small firms is considered as a privileged instrument of industrial policy and national planning; financing and marketing activities seek to organize the markets and the development of technologies through alliances, mergers, agreements and political interventions; the hope for an economic renewal is concentrated on small enterprises that are in full harmony with economic needs. A small enterprise is also a formidable machine able to enrich or destroy capital. It presents itself as a sort of carrier of values to the extent that it creates bridges for the transfer of productive resources (financial capital technologies, labour force with different qualifications and competencies) to activities, markets and large companies able to make profits.

The entrepreneur's individual qualities and personality undoubtedly play a major role in the decision to create or buy out a small firm. Nevertheless, the action of starting up that initiative is determined by the macro-systemic dynamics of accumulation and profit. These dynamics generate barriers as well as personal enrichment opportunities that encourage an individual to become an entrepreneur who will ultimately succeed or fail. The fact is that nobody is born an entrepreneur but may become one through the mobilization of a pool or resources composed of capital, knowledge and relations (Boutillier and Uzunidis 2001, 2006). Support involves capital for investments and operations, knowledge for choices and decisions, and relations for the financing, association and selling of products.

Therefore, the establishment of an enterprise is the result of the emergence of a defect in the economic structure caused by: (a) a profit differential due to the imperfection of markets and obstacles to the mobility of productive resources and goods; (b) the institutional mechanisms promoting and giving value to the entrepreneur's function; (c) the mobilization of necessary competencies and capital depending on that function. Thus, the contemporary entrepreneur and his/her small enterprise are indispensable for: (a) the consistency between the entrepreneurial areas of large enterprises; (b) the re-activation of the local micro-social fabric; (c) the timely amelioration of the burden of unemployment, inactivity and precariousness. For more than two decades, public authorities and big companies have implemented a formidable system in order to manage the creation of enterprises (with a view to increasing the entrepreneur's 'resource potential') and fill the gaps generated by economic activity as well as to avail themselves of the means necessary to experiment with new merchant values and social practices.

In the public administration system, the OECD (1999) urges governments and local authorities to intervene in the economy in order to 'foster the entrepreneurial spirit'. Financial logistic and legal aids are numerous. In the 1980s, public policies used to privilege innovative enterprises resulting from academic research (information technologies, microelectronics, new materials, biotechnologies, in particular pharmacology). The Silicon Valley model in California was largely copied in Western and Eastern Europe and other enterprise zones emerged around the major academic cities. The purpose is to commercialize research by introducing entrepreneurship into the activities of academics or engineers. This policy was also viewed as a measure targeted at regional development and planning, leading to the emergence of innovation clusters and poles of excellence through the mass creation of technological enterprises. However, policy has evolved since the 1990s. Aid given to the creation of enterprises must also satisfy social objectives, e.g. in relation to unemployment. Political leaders, together with banking institutions, are presently promoting the creation of enterprises by people who find themselves in difficult social and economic conditions. Today in Europe and in the United States, one enterprise out of two is created by an unemployed person or job-seeker.

For example, the French law of 1999 on research and innovation gave universities a new mission in addition to training and the production of knowledge: the commercialization of research. The purpose is to build bridges between the rather closed realms of research and business. However, in order that these links be translated into positive action for both universities (new research programmes for university labs) and regions (technology transfer, job creation) legislative measures must obviously facilitate the creation of such links; nevertheless academic strategies must be in phase with that specific project (raising the researchers' interest vis-à-vis the issue of commercialization). The scientific and technological level must be sufficient and adapted to the purpose of commercialization (or applied research). The economic and social environment must be compatible with the purpose of valuing research. In other words, there must be both incentives to create innovative enterprises and demands coming from the already existing industrial structure (enterprises) in order to absorb and distribute the results fielded by science and technology, as explained earlier when we described the different economic and structural measures related to the promotion of R&D investments (see Figure 8.2).

The most striking example relates to the implementation of a structural policy in order to assist entrepreneurs and increase the economic attractiveness of some milieu: the creation of 'competitive clusters' (Fontagné and Lorenzi 2005) all over the industrialized world but also in some major developing countries (India, China, etc.). In France, DATAR started in 1998 to list and to promote local production systems and industrial districts

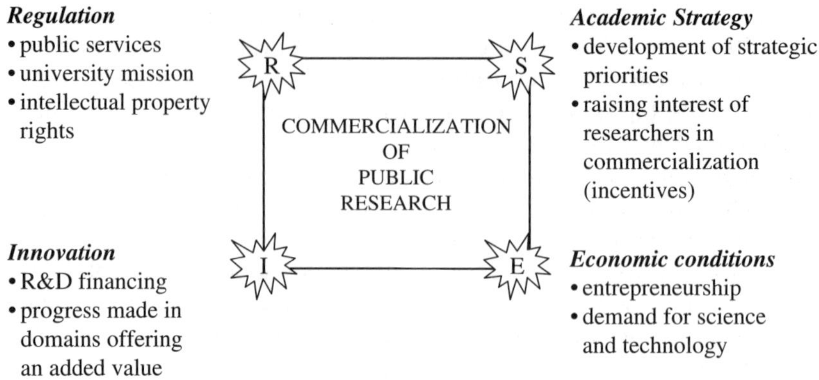

Source: Laperche (2002).

Figure 8.2 Key factors for commercializing research

in order to transform them quickly into competitive clusters. According to the official definition,[2] a competitive cluster is:

> the combination on a given geographical space of enterprises, training centres and public or private research units involved in a partnership approach with the purpose of identifying synergies around common projects of an innovative nature. This partnership is organised around a market and the scientific and technological domain connected to it; its purpose is to find a critical mass capable of ensuring international competitiveness but also worldwide visibility.

Competitive clusters combine both attractiveness (agglomeration effect) and diffusion (dispersion effect): these two factors are essential for the generation and propagation of innovations. What is favourable to an innovative entrepreneurship offering the possibility to support 'network economies' is the existence of an area created, in economic terms, by the following factors: a significant innovation potential, a diversified and wealthy system guaranteeing the protection of intellectual property and confidentiality, a dense network of SMEs, access to a large regional/domestic market, political incentives for research and innovation.

The assistance given to the creation of enterprises is part of both employment policy and industrial/innovation policy. A small enterprise is therefore 'manufactured' by the combined action of governments and by industrial and financial enterprises. This is clearly underlined by the Small Business Association (SBA) in the USA: in order to be competitive in international markets, a large enterprise must transform its internal functions into independent units and very often call upon specialized small enterprises. On the one hand, this process of activity outsourcing and streamlining (simplification of group organization charts) is facilitated by public legal and fiscal measures: legislation related to tax relief on holdings and the extra-territoriality of financial subsidiaries enable groups to manage their partnership contracts, sub-contracting agreements and licences more easily. The financial control required by splitting up production structures encourages the creation of small enterprises flourishing in market and specific technological 'niches'.

After having concentrated its production means, defined and divided its production tasks and created easily controllable corporate units, the large enterprise is now becoming a de-concentrated centre devoted to the organization and management of its productive resources. Nowadays, production is organized as if the power of an enterprise in the market (and the co-ordination of the functions and activities it may impose at that level) constituted a factor of economic power (and of centralization of asset property) much more important than the power conferred on it by its own assets (scientific technological, industrial and financial). The power exerted by the firm in the market results from its financial capacity (financial assets and mobilization of capital) and from its information capacity. The combination between information and finance is used to create and manage the corporate identity of small units that are geographically dispersed and physically distant (investments in inter-industry cooperation relations, protection of technological assets, acquisition of scientific knowledge and design of new goods, coordination through data communication services concerning various activities, etc.). Therefore, managerial coordination strengthens the manager's role in industrial organization and subjects him to the decisions of the managerial authority.

Small innovative enterprises become part of the networks established by major groups and are often co-ordinated with territorial development (Uzunidis and Boutillier 2003), e.g. scientific parks where companies endowed with broad technological and commercial advantages themselves create small specialized enterprises (similar to research labs) managed by academic entrepreneurs. In other cases, the middle management of a large enterprise is asked by the parent society to create a firm in order to test new technologies (intra-entrepreneurship); quite often we note the emergence of 'speculative entrepreneurship' (Tidd et al. 2005) serving large companies. For example, highly qualified persons, supported by preferential loans or public subsidies, may create an enterprise in order to sell it to a bigger unit once the product or process has been developed. The connection between small specialized enterprises and groups is achieved through an adapted financial and monitoring strategy. Venture capital societies (participation in the capital of a recently established firm, business angels (wealthy individuals investing in innovative projects) and other financiers (pension funds are very active) invest the capital of savers in new enterprises operating in the information and biotechnological fields.

3. CONCLUSION

Both the local and entrepreneurial aspects of innovation reveal the mechanisms for forming and acquiring all the ingredients (capital) of productive activity. The issue of appropriability has become crucial for the functioning (location or creation) of an enterprise. Enterprises tend to make greater use of their environment than to invest, e.g. in all the phases of technological creation. This may be explained by the fact that investments devoted to the acquisition (appropriation) of productive resources are cheaper than those made for the creation of these resources. This also explains the attractiveness (in an open economy) of a region that has abundant scientific and technological resources. The creation of innovative or more traditional enterprises is determined by the degree of wealth of the milieu. If the factors related to education, environment, medical care, financing infrastructures, housing, etc., influence the evolution of the marginal costs of an enterprise or activity, they also impact on the performance of the capital invested. In this case, the concepts of 'networking' and 'innovative environment' become predominant in our observations and economic analysis.

The achievement of innovation networks based on a review of synergistic relations inside the scientific and technological system of a country follows four pathways: creation of institutions for the commercialization of research; support for the creation of technological enterprises; local and regional planning starting from scientific and technological 'poles of excellence' or competitive clusters. It is the milieu that gives value to a research activity, enables the achievement of innovations, creates enterprises, and contributes to reassuring managers and entrepreneurs. The entrepreneur or enterprise that appropriates (and gives value to) all kinds of scientific and technological knowledge automatically plays a central role in the whole set of innovation and profit interrelations and opportunities.

NOTES

1. For a deeper analysis of 'innovation systems' and 'innovative milieux', see the special issue of the journal *Innovations, Cahiers d'économie de l'innovation* (2002).
2. Government website, http://www.competitive.gouv.fr, accessed March 2007.

BIBLIOGRAPHY

Abdelmalki, J. and J. Perrat (eds) (1992), *Les Nouvelles logiques du développement – globalisation versus localisation*, Paris: L'Harmattan.

Acs Z.J. and D.B. Audretsch (1990), *Innovation and Small Firms*, Cambridge, MA: MIT Press.

Agence française pour les investissements internationaux (AFII) (2005), *Le Tableau de bord de l'attractivité de la France*, Paris: AFII.

Aghion, P. and P. Howitt (1998), *Endogenous Growth Theory*, Cambridge, MA: MIT Press.

Albertini, S. (1997), 'Industrial district towards networks in the North-East of Italy: the new division of labour in the learning process', in C. Palloix and Y. Rizopoulos (eds), *Firmes et économie industrielle*, Economie et Innovation Series, Paris: L'Harmattan, pp. 239–54.

Amable, B., R. Barré and R. Boyer (1997), *Les Systèmes nationaux d'innovation à l'ère de la globalisation*, Paris: Economica.

Aydalot, P. (1986), *Les milieux innovateurs en Europe*, Paris: Economica.

Becattini, G. (1990), 'The Marshallian industrial district as a socio-economic notion', in F. Pyke, G. Becattini and W. Sengenberger (eds), *Industrial District and Inter-firm Cooperation in Italy*, Geneva: ILO.

Beffa, J.L. (2005), 'Pour une nouvelle politique industrielle', report to the Président of the Republic, Paris.

Benko, G., M. Dunford and A. Lipietz (1996), 'Les districts industriels revisités', in B. Pecqueur (ed.), *Dynamiques territoriales et mutations économiques*, Paris: L'Harmattan, pp. 119–34.

Betbèze, J.P. (2005), *Financer la R&D*, Conseil d'analyse économique, Paris: La documentation française.

Boutillier, S. and D. Uzunidis (2001), 'L'utilité marginale de l'entrepreneur', *Innovations, Cahiers d'Economie de l'Innovation*, **13** (1), 17–42.

Boutillier, S. and D. Uzunidis (2006), *L'aventure des entrepreneurs*, Paris: Studyrama.

Branscomb, L. and J. Keller (1998), 'Towards a research and innovation policy', in L. Branscomb and J. Keller (eds), *Investing in Innovation Creating a Research and Innovation Policy that Works*, Cambridge, MA: MIT Press, pp. 462–96.

Coase, R. (1937), 'The nature of the firm', *Economica*, **16** (4), 386–405.

Courlet, C. (1999), 'Territoire et développement', *Revue d'Economie Régionale et Urbaine*, **3**, 533–46.

Courlet, C. and B. Pecqueur (1992), 'Les systèmes industriels localisés en France: un nouveau modèle de développement', in G. Benko and A. Lipietz (eds), *Les régions qui gagnent, districts et réseaux: les nouveaux paradigmes de la géographie économique*, Paris: PUF, pp. 81–102.

Crevoisier, O. (2000), 'L'approche par les milieux innovateurs: état des lieux et perspectives', *Revue d'Economie Régionale et Urbaine*, **1**, 153–66.

De Bernady, M. (2000), 'Système local d'innovation: facteurs de cohésion et de pérennité', *Revue d'Economie Régionale et Urbaine*, **2**, 265–80.

Dosi, G., C. Freeman, R. Nelson, G. Silverberg and L. Soete (eds) (1988), *Technical Change and Economic Theory*, London: Pinter.

Dupuis, C. and J.P. Gilly (1996), 'Apprentissage organisationnel et dynamiques territoriales: une nouvelle approche des rapports entre groupes industriels et systèmes locaux d'innovation', in B. Pecqueur (ed.), *Dynamiques territoriales et mutations économiques*, Paris: L'Harmattan, pp. 157–75.

Fontagné, L. and J.H. Lorenzi (2005), *Désindustrialisation, délocalisations*, Conseil d'analyse économique, Paris: La documentation française.

Gaffard, J.L. (2005), 'Vers une nouvelle politique industrielle', *Lettre de l'OFCE*, **269**, 1–7.

Garofoli, G. (1993), 'Economic development, organization of production and territory', *Revue d'Economie Industrielle*, **64**, 22–37.

Gilly, J.P. and M. Grossetti (1993), 'Organisation, individus et territoires. Le cas des systèmes locaux d'innovation', *Revue d'Economie Régionale et Urbaine*, **3**, 449–68.

Gilly, J.P. and A. Torre (2000), *Dynamiques de proximité*, Paris: L'Harmattan.

Guellec, D. and P. Ralle (1995), *Les Nouvelles théories de la croissance*, Paris: Repères, La découverte.

Héraud, J.A. (2003), 'Régions et innovation', in P. Mustar and H. Penan (eds), *Encyclopédie de l'innovation*, Paris: Economica.

Innovations, Cahiers d'économie de l'innovation (2002), 'Géo-économie de l'innovation', special issue, **16**.

Konstadakopoulos, D. (2000), 'Milieux innovateurs et apprentissage dans le Sud-Ouest de l'Angleterre', *Innovations, Cahiers d'économie de l'innovation*, **11**, 139–54.

Krugman, P. (1993), 'First nature, second nature and metropolitan location', *Journal of Regional Science*, **33**, 129–44.

Laperche, B. (1998), *La Firme et l'Information*, Paris: L'Harmattan.

Laperche, B. (2002), 'The four key factors for commercialising research. The case of a young university in a region in crisis', *Higher Education Management and Policy*, **14** (3), 149–71.

Laperche, B. (ed.) (2003), *L'innovation orchestrée. Risque et Organisation*, Paris: L'Harmattan.

Laperche, B. and D. Uzunidis (eds) (2005), *John Kenneth Galbraith and the Future of Economics*, London: Palgrave Macmillan.

Laperche, B., J.K. Galbraith and D. Uzunidis (eds) (2006), *Innovation, Evolution and Economic Change. New Ideas in the Tradition of Galbraith*, Cheltenham, UK and Northampton, MA, USA: Edward Elgar.

Lecoq, B. (1995), 'La relation technologie-territoire et les milieux innovateurs', *Revue Internationale PME*, **8** (1), 81–106.

Madeuf, B., G. Carré, G. Lefebvre and C. Milelli (2005), 'TIC et économie de la proximité: organisation et localisation de la R&D au sein des entreprises globales', *Innovations, Cahiers d'économie de l'innovation*, **21** (2), 243–72.

Maillat, D. and J.C. Perrin (eds) (1992), *Entreprises innovatrices et développement territorial, Enquête GREMI II*, Neuchâtel: EDES.

Marshall, A. (1919), *Industry and Trade*, London: Macmillan.

Ministère de l'Education Nationale et de la recherche (2005), *Fiches régionales: les principaux indicateurs de R&D*, accessed May 2006 at http://cisad.adc.education.fr/reperes/public/nouveau/default.htm.

Mustar, P. and H. Penan (2003), *Encyclopédie de l'innovation*, Paris: Economica.

Nooteboom, B. (2002), A cognitive theory of the firm, *paper for ESNIE workshop Alternative Theories of the Firm*, accessed May 2005 at http://esnie.u-paris10.fr/pdf/ws_2002/Nooteboom.pdf.

Observatoire des sciences et des techniques (2004), *Indicateurs de Sciences et de Technologies*, Paris: Economica.

OECD (1999), *Stimuler l'esprit d'entreprise*, Paris: OECD.

Perrin, J.C. (1991), 'Réseaux d'innovation, milieux innovateurs et développement territorial', *Revue d'Economie Régionale et Urbaine*, **3/4**, 343–74.

Peyroni, J. (2002), *Le Schéma de développement de l'espace communautaire*, Paris: La documentation française.

Pitelis, C., R. Sugden and J.R. Wilson (2005), *Clusters and Globalisation*, Cheltenham, UK and Northampton, MA, USA: Edward Elgar.

Proulx, M.U. (1994), 'Milieux innovateurs: concept et application', *Revue Internationale PME*, **7** (1), 63–84.

Rallet, A. (1993), 'Choix de proximité et processus d'innovation technologique', *Revue d'Economie Régionale et Urbaine*, **3**, 365–86.

Rallet, A. and A. Torre (2002), 'Proximité Géographique ou Proximité Organisationnelle? Une analyse spatiale des coopérations technologiques dans les réseaux localisés d'innovation', *Economie Appliquée*, **1**, 147–72.

Robinson, J. (1972), *Hérésies économiques*, translated by G. Grellet, Paris: Calmann-Lévy, originally published by Basic Books.

Sierra, C. (1997), 'Proximité(s), interactions technologiques et territoriales', *Revue d'économie industrielle*, **82**, 7–38.

Small Business Administration (2000), *The Third Millennium: Small Business and Entrepreneurship in the 21st Century*, Washington, DC: SBA.

Tidd, J., J. Bessant and K. Pavitt (2005), *Managing Innovation*, Chichester: J. Wiley and Sons.

Uzunidis, D. (2003), 'Les facteurs actuels qui font de la science une force productive au service du capital', *Innovations, Cahiers d'Economie de l'Innovation*, **17** (1), 51–78.

Uzunidis, D. (ed.) (2004), *L'innovation et l'économie contemporaine*, Brussels: De Boeck.

Williamson, O.E. (1985), *The Economic Institutions of Capitalism*, New York: Free Press.

9. The entrepreneur's 'resource potential', innovation and networks

Sophie Boutillier, Blandine Laperche and Dimitri Uzunidis

1. INTRODUCTION

In the current so-called 'knowledge economy', innovation is considered as the engine of economic growth (Romer 1990; Aghion and Howitt 1998; Audretsch 2006). For most of the second half of the 20th century the large firm was considered as being at the origin of innovation (thanks to its ability to gather large amounts of resources). However, since the 1980s, the entrepreneur has made his/her comeback at the forefront of the economic and political scene. What is the exact role of the entrepreneur and how does he/she succeed in the innovation process, which still – and perhaps, because of its rapid pace, more than ever – requires the gathering of large amounts of finance, competencies and information? To answer this question, we firstly and principally refer to the work of several economists, especially J.A. Schumpeter (1883-1950), who can be considered as a precursor in the analysis of entrepreneurship linked to the innovation process.

The entrepreneur is the one who disturbs the neoclassical equilibrium by executing new combinations in the means of production (Schumpeter [1911] 2006), who puts the economy on the path of motion and development (Schumpeter 1939), but also the one who, by his extinction, accounts for the self-destruction of capitalism (Schumpeter [1942] 1975). The entrepreneur, the first hero of economists according to Schumpeter, is incontestably Schumpeter's own hero. But the hero is an elusive one: being an entrepreneur is not a profession but a function according to Schumpeter, and thus entrepreneurs have a butterfly-like existence.

What is the origin of the entrepreneur's function? Is entrepreneurship a natural gift, as Schumpeter seemed to believe? Why does the entrepreneur's function disappear and is not long-lasting? In an approach combining economics and sociology, we will try to go deeper into the analysis of the social origin of the entrepreneur's function, studying the construction of what we call his[1] 'resource potential', i.e. the set of knowledge, relations and financial resources gathered together by the entrepreneur in his environment. This resource potential is not stable and may be increased or reduced, in different economic, political and social contexts. However, this chapter does not try to define the exact ingredients of a 'good' public policy for entrepreneurship. As a matter of fact, since J.M. Keynes on the one hand and also the work of the economists of the Public Choice School on the other, we have learnt that the impacts of public policies do not always give the expected results (Keynes 1933; Buchanan and Tullock 1962). Our aim consists more in emphasizing the entrepreneur's socio-economic background. Notwithstanding his own qualities (smartness, ability to judge), the entrepreneur is for us a social actor, influenced by the social, economic and political context. This context defines several socio-economic functions; particularly in our case, the entrepreneur's function (ownership of means of production, individual initiative, business creation, etc.), which takes form through the genesis and the diffusion of major or incremental innovations. The analysis in terms of 'resource potential' aims at going deeper into the explanation of what turns the potential entrepreneur into a real one.

In the first section of this chapter, we will come back to socio-economic and historical analysis of the entrepreneur's function, showing that this character is concurrently the driving force of capitalism, as well as the reason for the evolutionist character and the questionable self-destruction of this economic system. In the second part, we will point out the origins of the entrepreneur's function and develop our analysis of the resource potential in relation to a network approach, which is fed but also differentiated from current – or more ancient – approaches in terms of social capital. The formalization in terms of an 'organic square of entrepreneurship' which we propose, is thus an attempt to explain the temporary character of the entrepreneur's function. It also supports our thesis according to which today's entrepreneur can be sketched as a socialized entrepreneur (a network-based entrepreneur), at the interplay between the strategies of states and large firms.

2. BUSINESS CREATION AND THE DYNAMICS OF CAPITALISM

Schumpeter was not the first economist to draw attention to the entrepreneur. Richard Cantillon (1680–1733), Jean-Baptiste Say (1767–1832), Karl Marx (1818–83) and Alfred Marshall (1842–1924) all made important contributions to the economic analysis of this character (Boutillier and Uzunidis 1995, 1999 and 2006a). However, Schumpeter's approach is original in that he formalized the entrepreneur's function in relation to the dynamics of capitalism (Uzunidis 2006a).

2.1 The Entrepreneur: the Driving Force of Capitalism?

First of all, the entrepreneur, as viewed by Schumpeter (1911), is the one who questions the economic status quo prevailing in the neoclassical vision. In fact, for neoclassical economists, and especially their leader Léon Walras (1834-1910), the basic model is the one of pure and perfect competition. Competition is the only way the market can operate. In this context, there exist neither big firms nor small firms, but only firms of similar size (the assumption of atomistic markets). Firms are and will remain of similar size, which accounts for the absence of technical progress. The economy duplicates itself indefinitely. For Walras, the industrial entrepreneur buys textiles and raw materials, rents factories, workshops, machinery and tools, hires spinners, blacksmiths and mechanics (Walras [1954] 1988, p. 287). The Walrasian entrepreneur does not innovate: he buys or rents services or raw materials or other input required by production, and selects according to the price of factors of production (interest rate for capital, wage rate for labour).

The Schumpeterian entrepreneur questions the economic status quo because he is the economic agent that makes new combinations of productive factors corresponding to as many investment opportunities. He is the driving force of the capitalist economy, embodying the dynamics of capitalist change as well as industrial and technological progress. There are five such combinations:

> the introduction of a new good – that is one with which consumers are not yet familiar – or of a new quality of a good.
>
> the introduction of a new method of production, that is one not yet tested by experience in the branch of manufacture concerned, which need by no means be founded upon a discovery scientifically new, and can also exist in a new way of handling a commodity commercially.
>
> the opening of a new market, that is a market into which the particular branch of manufacture of the country in question has not previously entered, whether or not this market has existed before.
>
> the conquest of a new source of supply of raw materials or half-manufactured goods, again irrespective of whether this source already exists or whether it has first to be created.
>
> the carrying out of the new organisation of any industry, like the creation of a monopoly position (for example through trustification) or the breaking up of a monopoly position. (Schumpeter 1911/2006, p. 66)

According to Schumpeter, to be an entrepreneur is neither a profession nor a long-lasting state. On the contrary, to be an entrepreneur is a 'function'. This is the second characteristic of the Schumpeterian entrepreneur: 'the entrepreneur's function is to combine the productive factors, to bring them together' (ibid., p. 76). The entrepreneur is then very different from the common vision of the manager, who manages the day-to-day production: 'everyone is an entrepreneur only when he actually "carries out new combinations", and loses that character as soon as he has built up his business, when he settles down to running it as other people run their businesses' (ibid., p. 78). To be an entrepreneur does not automatically mean having a long-lasting relation with an individual concern, as stressed by Say or Marshall in their own words.

For Say, the entrepreneur stands between the operative work of the worker and the research work of the scientist 'L'entrepreneur met à profit les facultés les plus élevées et les plus humbles de l'humanité. Il reçoit les directions du savant et les transmet à l'ouvrier'[2] (Say [1820] 1999, ch. VI, p. 96). As a result, there is a distinction between three kinds of industrial operations: the research by the scientist, the application by the entrepreneur and the execution by the worker. However, according to Say, the entrepreneur may at the same time lead research, apply the results to industry and himself manufacture what will be sold. In volume 1 of *Principles of Economics*, Marshall defined the entrepreneur as putting forward his managerial skills, his faculty of knowing how to organize the work of an important number of people. His function is to manage the production in a way that a given effort may have the most important effect on satisfying human needs (Marshall, 1890).

The entrepreneur can then have various profiles: he can either be independent or the employee of a public company, or also have no durable relation with an individual business. As a result, Schumpeter had a particular view of the connection between the entrepreneur, risk and profit. One of the meanings of the verb 'to risk' being 'to attempt' or 'to undertake', the entrepreneur and risk are directly linked, as shown by the history of economic thought referring to the entrepreneur (Boutillier and Uzunidis 1995, 1999; Blaug 1998; Laperche 2003). The first set of key issues refer to the nature of the entrepreneur's remuneration. Then, as the dissociation between the entrepreneur and the capitalist becomes clearer, economists will be interested in the entrepreneur's function and the associated risks.

In classical economic thinking, and particularly in the works of Adam Smith (1723–90), the economic agents that are the entrepreneur and the capitalist were often confused, as well as the financial risk and the undertaking risk. Before Smith, Richard Cantillon, who was, according to Schumpeter ([1954] 1983, p. 242), the first economist to use the term 'entrepreneur', saw the entrepreneur's remuneration as a fair remuneration of the incurred risk. According to him, entrepreneurs (the term is used, taking account of the period, to qualify merchants) are 'people with uncertain revenues' because their revenues depend on the risk they take in a context of uncertainty. Their function consists of acquiring means of production at certain prices with the purpose to sell at uncertain prices. For Schumpeter, the entrepreneur's role does not only consist of taking economic risks. Entrepreneur and capitalist are not synonymous. The ordinary shareholder is not always an entrepreneur. The capitalist is only an owner of money, debts or goods.

Say, in the context of industrialisation, and Schumpeter, when the movement of capital concentration gained ground, gave substance to the notion of risk, as they particularly focused on the entrepreneur's function. For these two authors, even if they express it differently, the risk undertaken is the risk linked to innovation. This point of view was also the one of Ricardo and Marx, who considered that industrial gains were composed in part of the remuneration they receive, for a time from the first introduction in the economic process of an original improvement, as for example a new machine. They thus discovered, according to Schumpeter, a special case of the most typical gain of the entrepreneur (Schumpeter [1954] 1983, pp. 361–2). However, while Say and Marshall – who also draw attention to the organisational and productive function of the entrepreneur – tried to justify the profits of the entrepreneur, Schumpeter asserted that profit is not the entrepreneur's first motivation. His behaviour is similar to the gambler. The entrepreneur is prompted by a desire for power. The joy to win and to create prevails in the intrinsic lure of gain. Profit crowns the success of the 'new combinations'. This is the expression of the value of the entrepreneur's contribution to production in the same way as the wage is the reflection of the worker's production.

Finally, apart from these psychological characteristics of the entrepreneur, Schumpeter heavily stressed the capacity of this entrepreneur to overcome the banker's resistance, the latter feeling an aversion for risk linked to innovation. According to Schumpeter, the existence of the entrepreneur is not sufficient to put the economy on the path of evolution. The potential entrepreneur needs access to credit, i.e. the admission ticket to participate in the game: 'The requisite funds – his admission ticket to the social store of means of production – he borrows from a bank' (Schumpeter 1939, vol. 1, p. 131; see also Ulgen 1996). In *Capitalism, Socialism and Democracy* (1942/1975), Schumpeter explained that credit is at the origin of development, saying that it is the source on which one can draw to finance the execution of new combinations.

2.2 The Entrepreneur, Innovation and Progress

In *Business Cycles* (1939), Schumpeter analysed the cyclical process of evolution of the economy. This stems from the business spirit, the entrepreneur's activity and the introduction of radical innovations.

Innovation, in the broad sense of the word, is for Schumpeter the only origin of competitiveness, the determining element of accumulation. It is also at the origin of disequilibrium and imperfections. Schumpeter extends the marginalist theory (but then questions it) and cleverly explains, through the concept of innovation, the emergence of profit differentials at a given time in a given economy driven by pure and perfect competition as well as the evolution of economic activity.

Consequently, the role of innovation – and thus of the extraordinary person, the entrepreneur – is essential to the explanation of economic cycles (Uzunidis 1996, 2006b). According to Schumpeter, since only the most enterprising and venturesome people act, innovation appears in 'swarms'. These innovations induce a rise in the demand for capital goods, a decrease in production costs and an increase in the number of goods supplied. The increased capacity on the supply side is associated with a boom in demand stemming from the new consumption needs, the positive anticipations of economic agents and the development of bank loans. The ensuing rise in profits puts the economy on an ascending trajectory. This stage of prosperity, after the stimulating effect of radical innovation, is prolonged by minor innovations. The boom however limits itself as, paradoxically, innovation contributes to downsizing. Investment and excessive optimism during the years of strong growth may explain the cyclical shifts: losses, bankruptcies and job-cuts, and so on. The crisis will last as long as there exist production capacities in excess. When adjustment to the innovation is complete, deflation ends and the Walrasian equilibrium is restored. This thorough clean-up of the economy releases the business spirit so that a new innovation wave may occur.[3]

According to Schumpeter, but also to Marx, technical progress results in the increase in the scale of production, first through the development of the limited company: the limited company supplies capitalism with the necessary capital for the development of large-scale industry. The entrepreneur disappears when the large firm emerges (resting on the partition of organizational work and production work, which is basic to automation), managed by a double-headed team, the managers and the shareholders (the partition between management and ownership of capital): 'Technological progress is increasingly becoming the business of teams of trained specialists who turn out what is required and make it work in predictable ways' (Schumpeter [1942] 1975, p. 132).

The entrepreneur, who was the driving force of capitalism in the heroic age, has then been replaced by a team of specialists whose essential task is innovation. So capitalism becomes impersonal, and the organization has replaced the individual. Schumpeter thus shared with Marx the concept of self-destruction of capitalism. In *Capitalism, Socialism and Democracy*, Schumpeter explained that capitalism has made technical progress its *raison d'être*: 'Capitalism, then, is by nature a form or method of economic change and not only never is but never can be stationary' (ibid., p. 82). Is the entrepreneur not the driving force of capitalism just because he innovates, thus continuously creating new investment opportunities?

Marx wrote: 'the actual barrier to capitalist production is capital itself' (*Capital*, book III, [1894] 1976, p. 244, our translation). 'Capitalist production generates its own negation which conditions the transformations of nature. It is the negation of the negation' (*Capital*, book I, [1867] 1976 p. 557, our translation). Schumpeter wrote 'that those factors make not only for the destruction of the capitalist but for the emergence of a socialist civilization. They all point in that direction. The capitalist process not only destroys its own institutional framework but it also creates the conditions for another' (Schumpeter [1942] 1975, p. 162). Marx and Schumpeter shared the idea of self-destruction of capitalism, that self-destruction being paradoxically caused by its success. For the latter, the development of shareholders have destabilizing effects on the future of capitalism, thus joining Adam Smith for whom the development of joint-stock companies was incompatible with the development of business, as the private appropriation feeling is absent. It is dissolved in the multiple ownership of capital. For Schumpeter, the occurrence of collective private property destroys initiative and the will to enrich oneself which are the cornerstones of capitalism. The spirit of enterprise vanishes, and the firm is no longer a place of economic opportunities, nor a place of professional and personal fulfilment, all that being replaced by a bureaucratic organization.

> The capitalist process, by substituting a mere parcel of shares for the walls of and the machines in a factory, takes the life out of the idea of property. It loosens the grip that once was so strong – the grip in the sense of the legal right and the actual ability to do as one pleases with one's own; the grip also in the sense that the holder of the title loses the will to fight, economically, physically, politically, for 'his' factory and his control over it, to die if necessary on its steps. (ibid., p. 142)

2.3 Networks and the Socialized Entrepreneur

However, in this early stage of the 21st century, not only has capitalism not disappeared but its power over global society is reinforced. Large multinational firms, TNCs, dominate the world economy, while myriads of small firms are born, grow and die on a regular basis, thus composing a kind of stock from which capitalism finds its means of development. 'The age of the Entrepreneur is like Camelot: it is only here for a brief, shining moment each (roughly) half century', declares Ray Canterbery (2001, p. 423).

Therefore the domination of big firms is not incompatible with the maintenance of large numbers of small firms. Quite the opposite! The strong economic growth that came after World War II had provided Western Europe and the United States with safe and stable markets. After the trauma of the 1929 crisis, which had paved the way for war and dictatorship, all-out state intervention had become necessary (support for supply and demand). The economic crisis starting in the 1970s questioned many certainties. The industries and activities on which economic growth was based had reached maturity. The information and communication technologies (ICTs) generated by military programmes represented new investment opportunities. Therefore it became necessary to release the capital which had been frozen by nationalizations. This was done by financial revolution. The limited company had a free hand. It developed through venture capital and pension funds. This was the way 'new capitalism' saw the light. The division of labour had become more complex within and between firms. Firms communicated in real time and all over the world on the Internet. Financial markets expanded on a global scale. The gap between capital management and property widened. It mattered little whether the ownership of capital was collective; what was important was the individual possession of the profits that capital generated.

The economic history since the 1980s confirms this observation: domination by big firms, and survival of myriads of small companies. The theory of corporate governance (even if it reverses the balance of power between managers and shareholders) is based on that concept of bureaucracy and collective organization. However, how can we account for the economists' craze for the entrepreneur? Has the latter become the driving force of capitalism again? Is (entrepreneurial) capitalism the end result of human evolution, as Fukuyama (1993) claimed?

There are, according to us, diverse reasons for the entrepreneur's comeback in economic theory. These include (see Boutillier 2006b):

1. *The slow-down in economic growth and the persistence of massive unemployment*. The formation of new firms is perceived as a way of creating wealth (and boosting technological innovation), but also of creating employment. The establishment of new firms is perceived by political leaders as a social integration facilitator.

2. *The decrease in government social spending, the privatization and deregulation of the economy*. Facilitating the formation of companies is less costly to public budgets than granting unemployed people a substitute income.
3. *The occurrence of ICTs, providing new investment opportunities*. Small firms (often affiliated with large firms through financial relations) test those new fields.
4. *The development of financial flows, following the liberalization of the economic forces and markets*. These offer new financial instruments which both feed and weaken economic growth due to their high instability.

The large firms which control global markets are led by organizations, and not by one individual who can easily be identified: the entrepreneur and the founder. The firm is led by managers on behalf of its shareholders (the owners of the firm), as John Kenneth Galbraith already explained in the late 1960s (Galbraith 1967). Apparently, the development of financial markets since the early 1980s has restored part of their lost power to the shareholders, as these now get rid of unsatisfactory managers. However, despite the questioning of the vertical integration of big firms in favour of a network structure (made possible by the development of ICTs), large firms are still the global market organizing entities. 'Corporate power lies with management – a bureaucracy in control of its task and its compensation', wrote Galbraith in his last book published in 2004 (Galbraith 2004, p. 31; see also the analysis of this debate in Laperche 2005; Laperche and Uzunidis 2005; Laperche et al. 2006). Since the 1960s, he has not departed by a hair's breadth from his position. The development of corporate bureaucracy has not resulted in the disappearance of capitalism, or the entrepreneur. The latter even receives assiduous care from the political leaders who have crowned him as a creator of innovation and employment, as well as the champion of social integration.

We thus can make the following assumption: the entrepreneur has not disappeared, but he is no longer a hero. He has become a socialized entrepreneur, an entrepreneur who is a player in the economy made up of: (a) the planning decisions made by big firms which, even if networked, nevertheless are powerful bureaucracies, and (b) state policies which aim at reducing unemployment and boosting innovation in the developing knowledge economy. This is the new capitalism of the early 21st century!

Faced with the complexity of today's economic activity, Castells (2001) went as far as to maintain, quite cleverly, that the fundamental unit of the economic system is no longer the entrepreneur itself, the family, the firm or the state, but the network composed of different organizations. As regards innovation, the division of labour and the very refined specialization of skills in the process of production remove any possibility of autarkical organization of technological production. Partnerships between companies and between public or non-profit institutions on the one hand and companies on the other, and a whole panel of technical, financial and commercial contributions, illustrate the theories of the classical economists (e.g. Smith and Marx) for whom, as fast as industry takes over social production (and enlarges its market by appropriating the resources at the time), technical transformations are to be witnessed: core changes in the organization of production and in the ways the entrepreneur conducts business. The entrepreneur is not a deus ex machina who puts the economy on the path of motion. In the following section, our aim is to explain the economic and social origins of the entrepreneur's function, through a discussion of the profile of the entrepreneur considered as an economic player in the contemporary network-based economy.

3. THE ENTREPRENEUR'S RESOURCE POTENTIAL AND ENTREPRENEURIAL DYNAMICS

3.1 The Resource Potential: the Origin of the Entrepreneur's Function

We have developed the concept of 'resource potential' in order to assess the role played by the social origin, the educational background, the professional experience and the financing sources of entrants (Boutillier and Uzunidis 2007; Boutillier et al. 2004). Each individual owns a set of resources, and uses his or her potential to improve their economic situation (to find a job with a good salary, to create an enterprise, to find another professional activity, etc.) and their choices also depend on their own resources. The resource potential is not a natural gift but it is the product of a social process, of the opportunities and constraints implied by this process. What are the entrepreneurs' social origins? What did they study? Who were their parents? Were they entrepreneurs, or employees or civil servants? What are their relations?

The resource potential can be analysed in three respects: (1) knowledge (schooling, secondary education, higher education, further education and professional experience, etc.); (2) financial resources (personal savings, bank loans, venture capital and different forms of public support); (3) relations (family, personal, professional, institutional relations, etc.) (Table 9.1).

These three aspects are interdependent. For example, the knowledge of an individual depends on his education and/or professional experience. And the family context positively or negatively influences the educational

Table 9.1 The entrepreneur's resource potential: elements of definition

Resource potential	Main characteristics
Knowledge	• Tacit knowledge obtained from the family • Scientific and technical knowledge learned at school • Knowledge obtained through relations • Knowledge obtained through professional experience
Financial resources	• Personal savings • Love money: parents, friends, etc. • Bank loan • Financial aid from institutions (e.g. public aid) • Seed money from another entrepreneur
Relations	• Informal relations (family, friends, neighbourhood, colleagues, etc.) • Formal relations (state, banks, other enterprises, research institutes, etc.)

choices (Casson 1990). Many entrepreneurs were born to families of entrepreneurs. While studying, the individual meets others who can become partners and bring knowledge, money and relations. Making up a substantial address book is fundamental because it facilitates the search for money, markets and associates, etc.

Schumpeter considered that the entrepreneur's function was the result of a kind of 'not explained ability' and that his success was due to his competencies and also some luck: 'in most cases the man who rises first *into* the business class and then *within* it is also an able businessman and he is likely to rise exactly as far as his ability goes – simply because in that schema rising to a position and doing well in it generally is not or was one and the same thing' (Schumpeter [1942] 1975, p. 74).

Our approach therefore goes deeper into the analysis of the origin of the entrepreneur's function, trying to add to the notion of 'ability', a social dimension. Of course, according to Schumpeter, the successful entrepreneur is the one who has the banker's support. This is perhaps part of his explanation of the entrepreneur's ability. But, if Schumpeter prioritized the entrepreneur's ability to raise funds, this ability is, according to us, part of a multidimensional ability in which knowledge and relations also have an important role to play. Our approach of the resource potential is derived but also differentiated from the current (or more ancient) approaches in terms of social capital, presented below.

3.2 The Theoretical Roots of 'Resource Potential': an Economic Approach to Social Capital

The concept of social capital officially appeared in human sciences at the beginning of the 1980s. Bourdieu (1980) was one of its main originators. His aim was to go deeper into the analysis of social relations. Becker (1975) had earlier defined the concept of 'human capital', considering that all individuals own their knowledge and that they can take the decision to invest in their human capital to increase their value in the labour market. With the concept of social capital, social relations become valuable resources (Boutillier 2006a). The concept of social capital is the foundation of the concept of resource potential. However, in the economic and sociological literature, social capital is most often defined as a set of social relations. According to us, social relations are only part of the resources that individuals can invest in to find a job, start a business, etc.

Capitalizing one's own knowledge

At the beginning of the 1960s, the concept of human capital became essential, in economic theory as well as reality. Enterprises, notably those that were large, needed more and more skilled employees, due to the development of service industries and a clearer separation between design and manufacturing, the affirmation of full employment, the development of the educational system, the continuous technical progress and the resulting automation of industrial work.

Becker's theoretical breakthrough in the 1960s had long been preceded by Adam Smith, who emphasized the necessary development of education. But the true precursors of 'human capital' were certainly T.W. Schulz (1959) and E.F. Denison (1962). The quality of human resources, the investment in education, the information of individuals, and knowledge are factors of production. Becker's human capital is different from physical capital or financial capital because the knowledge and know-how acquired by an individual during his training are embedded in him. The employer can take advantage of this knowledge if the employee keeps on working in his enterprise. Employers do not have a property right on their employees.

However, human capital is still a form of capital and as such the theory of capital can be applied to it, adapting it and putting forward some specific properties: human capital can be accumulated. All individuals may increase the number of years dedicated to its formation, normally through a quasi indefinite process. Marginal returns of investment in education fall: over the life cycle, the accumulation of human capital follows a decreasing trend. As years go by, economic agents dedicate less time to training, all the more so as job perspectives become less numerous. The human capital return decreases with age: the integration of human capital in the individual limits his investment capacity. He is dependent on his physical and intellectual capacities. Moreover, an individual may accumulate human capital, without any positive result (added value in the labour market); human capital is both a private and a public good as an educated population is also more productive (cf. GDP growth, Lucas 1988).

The theory of human capital therefore presents similarities with that of capital. The individual invests in himself, as the capitalist invests in his enterprise: the former is incorporated in the individual, but the capital remains exterior to the entrepreneur who invests it. The law of decreasing marginal returns applies to human and physical capital alike.

Capitalizing one's social relations

Since the early 1990s, various works have been conducted by American sociologists (e.g. Putman and Coleman) in this direction. Social relations are considered a resource which can be capitalized by individuals. In order to improve their economic welfare, individuals use their social relations which are regarded as an individual and a collective good. According to these authors, the characteristics of social capital are similar to those of physical capital: decreasing returns, decreasing marginal productivity, etc.

Pierre Bourdieu (Bourdieu and Wacquant 1992) defined the concept of social capital in an analysis of social stratification. Differences in terms of revenues are not the only criterion of differentiation between social classes. For equal revenues, social behaviour and notably cultural behaviour can be very different. This social capital is a means to gain access to jobs, funding etc., and the individuals belonging to dominant social classes are those who benefit from a wide network of social relations. Bourdieu defined social capital as the sum of actual or potential resources which are linked to the more or less institutionalized network of inter-acquaintance and inter-recognition relations possessed by an individual. Every individual is endowed with a social capital; however the latter is not a natural gift but the result of a strategy. Consciously or not, individuals invest to increase their social capital, this depending on other forms of capital: economic, cultural and symbolic capital. A high economic or cultural capital facilitates the formation of a social capital. And a high social capital is also a means to increase the economic capital of individuals.

Long before Bourdieu, other sociologists had drawn attention to the potential wealth that constitutes a network of social relations. Max Weber ([1904–1905] 1964) showed the link there was in the USA at the beginning of the 20th century between the adherence to a religious community and success in business. In 1963, Macaulay published the results of a study conducted in the USA on transactions between firms, which are very often agreed without contracts or in inaccurate legal conditions. Entrepreneurs are often reluctant to formalities which may question trust. For American sociologists who studied the concept at the end of the 1980s, social capital can also be considered as a network of various relations enabling the individual to improve his social insertion. Subtle links between social capital and human capital are also established.

The issue of social capital is only apparently simple (Gasse et al. 2004). It is in fact possible to find as many definitions as there are authors who draw attention to it. Coleman (1988, 1990) describes social capital as a resource incorporated in inter-individual relations. The resources drawn from social capital are privately appropriated and have certain similarities with externalities generated by the functioning of the market. Social capital is productive, just as human capital is, and contributes to the increase of welfare. Putman (2000) defines social capital as the network and reciprocal norms which are associated with it. Social capital has a given value for the individuals who are part of the network. Burt (1995) stressed the role of social networks in the evolution of carriers, studying the promotion of executives in a large technological firm.

International institutions, the OECD and the World Bank for example, were quickly convinced of the interest in social capital in order to improve efficiency of public policies. OECD credits social capital with a major part in the increase of social welfare, considering three forms of capital linked by institutional arrangements: natural and product capital, human capital and social capital. For the World Bank, social capital is also a means to fight against poverty in order to fill the gap created by the absence of institutions. Social capital also supports informal economic relations, which depend on subtle relations of mutual trust and solidarity. In terms of public policies, the translation lies in the development of the associative sector (World Bank 2000).

The concept of social capital also has certain similarities with that of the network, which was also introduced in social and human sciences in the 1990s. The image of the network as a sum of computer connections has led many researchers to develop a metaphoric analysis (Castells 2001). Moreover, this kind of approach is interesting in that it takes account of the historical dimension. Commercial, political and financial networks have always existed, but they have taken on a new dimension with technical progress. This approach is not compatible with methodological individualism. The network is not seen as an appropriable good, capitalized by its members. The approach in terms of networks is considered as a method to study the world's evolution. The network is also used in different contexts, like for example the analysis of social cohesion, as an intermediary level between micro and macro. The accent is on the inter-personal links used to explain macro sociological phenomena (Granovetter 1973, 1985).

In a nutshell, those approaches in terms of human capital and social capital are interesting in that they question the pure rationality of individuals, putting them right in the middle of society (notably the role of family and state are clear) or in the 'strategic group' (Porter 1982) they have chosen. In the management literature, the role of social relations and of networks is more and more used to explain the entrepreneurs' success (Sammut 1998; Davidsson and Honig 2003; Suire 2004; Marchesnay 2004; Chabaud and Ngijol 2005). But these approaches seem too simplistic to us, because they remain largely focused on the individual, with very few references to the economic, social and political context in which he acts. Our analysis based on the 'resource potential' can thus be considered as a further step in the social analysis of the requisites for an individual's success – more precisely the entrepreneur – or of the explanation of his failure.

3.3 The Organic Square of Entrepreneurship and the Network-based Entrepreneur's Function

The resource potential is not a natural gift but is built by a conscious (or unconscious) strategy, linked to the economic, political and social context in which the entrepreneur operates. Moreover this potential is not stable. It can be increased, as explained by the abovementioned American sociologist, but it can also decrease in some special cases or contexts. For example a bankruptcy in Europe is very badly perceived and may be a case of drastic reduction in the resource potential; in another context, it may not change anything or even be regarded as an entrepreneurial attempt, which has to be followed by other attempts. In the same vein, public policies may help to increase an entrepreneur's resource potential or cause its decrease (support for business creation, taxes connected with the creation of the enterprise, etc.). The economic network-based organization (regulation, financial system, place of large enterprises, level of technical and knowledge development) and the market (supply and demand) may have the same positive and negative results on the resource potential of the entrepreneur and more globally on entrepreneurship. Our formalization in terms of an 'organic square of entrepreneurship' is thus an attempt to explain why the entrepreneur's function is not long-lasting and stable. Being an entrepreneur is not a profession but a function, which stops when the entrepreneur does not innovate any more: the social, economic, and political context may cause the appearance or the disappearance of the entrepreneur.

First of all, the economic context is characterized by change. Change is created by the activity of entrepreneurs, and major change comes in a context of crisis. Change also creates new investment opportunities (new needs causing an increase in demand, and so on). The economic situation is thus fundamental to the emergence of the entrepreneurial function. Secondly, as mentioned earlier, even if the entrepreneur was for Schumpeter a kind of *deus ex machina* he also has some competencies, and notably the one of convincing the banker and thus raising funds. The financial resources are thus taken into account. More globally the level of development of the financial system (and more precisely the banking system) is therefore very important to explain the entrepreneur's existence and success. Thirdly, institutional change (which characterizes economic organization in our scheme) is at the origin of the extinction or the revival of the entrepreneur's function. Our attempt here is to formalize the elements of the economic social and political context which may explain why the entrepreneur's function is not a stable one.

As a matter of fact, the entrepreneur's dynamism and qualities cannot explain per se the creation and success of small enterprises. This is the reason why we propose a more systematic analysis of the key factors of entrepreneurship (or the network-based entrepreneur's function) – high-tech or not – in today's economy. Four major factors affect this: public policies, economic and social organization (the importance of large corporations, the nature of the financial system and the development of knowledge), and the state of supply and demand which contribute to the building of the resource potential. The interdependence between these factors allows us to analyse them through an 'organic square of entrepreneurship' (Figure 9.1).

Why do we propose the expression 'organic paradigm of entrepreneurship', and how can we explain the choice of the incentive factors proposed

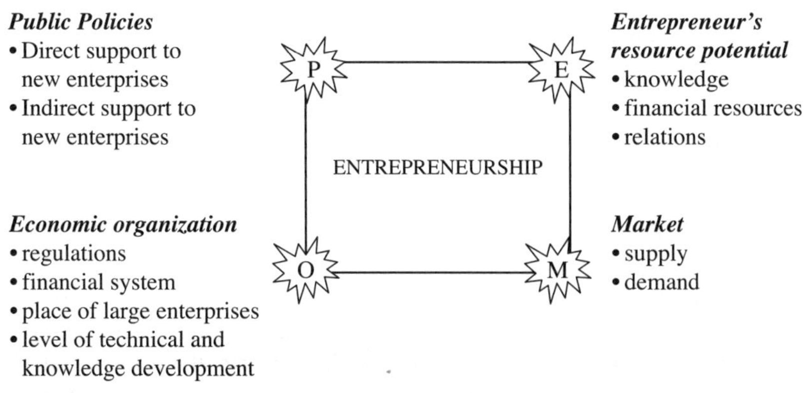

Figure 9.1 The organic square of entrepreneurship

here? To answer these questions, it is necessary to focus on the key factors of success of small businesses. Shane (2003) for example synthesizes the Schumpeterian entrepreneur (who is an innovator) and Kirzner's entrepreneur (who has to be well informed to catch the market opportunities, Kirzner 1973, 1985). Shane defines the relevant environmental factors (institutional, political and socio-cultural) which may help the entrepreneur to innovate. However, he does not make reference to the entrepreneur's own resource potential. Many other studies conducted by specialists in management sciences have demonstrated that the creation and durability of small enterprises depend on the entrepreneur's ability to take into account the existing links between different internal factors (organization, financial resources, activity) and external factors (the different levels of the environment) (Sammut 1998; Marchesnay 2003). In this case, the approach is more centred on the individual. Our analysis thus links the two approaches, focused on the entrepreneur and on the environmental factors.

This entrepreneurial 'maturity' results firstly from the variety and abundance of the resource potential which they have built. The composition of this potential depends on external factors and on the entrant him/herself. In particular, public support (direct or indirect) for the creation of new enterprises usually determines the financial resources entrepreneurs can have access to in order to set up or develop their business. The economic and social organization has several dimensions and different effects. The general level of development of knowledge influences the knowledge gathered by entrepreneurs (through their education and that of their staff, through economic intelligence) and the technological level of their activity. The characteristics of the financial system (possibility or difficulty of being listed on the stock exchange, more or less 'conservative' banks, and access to venture capital) have an impact on the capacity of the individual to become an entrepreneur and on the development of new businesses. Market concentration and the place of large companies also influence the dynamism of entrepreneurial activities and the kind of activity they have. Finally, the overall economic situation determines the rhythm of new creations and also the types of activities conducted by such businesses (see Boutillier et al. 2007).

It is in this organic square of entrepreneurship that the entrepreneur's function lies. This is a temporary function because it stems from the gathering at one moment of some special conditions. According to the context, it is also possible to examine different types of entrepreneurs with different resource potentials – e.g. high-tech and/or low-tech entrepreneurs. Our approach thus permits going deeper into the analysis of the entrepreneur, no longer as a disembodied agent but a social agent, in all its complexity.

4. CONCLUSION

In our opinion, explaining the origin (resource potential) and the conditions of appearance of entrepreneurs (the organic square of entrepreneurship or network-based entrepreneur's function) is all the more important as the entrepreneur has not disappeared, contrary to Schumpeter's prediction. He is at the forefront of the economic scene. The entrepreneur is also in the heart of the political debate, his existence and durability (as an economic act, and not as a function) being considered as the reason for economic growth. However, as explained above, our hypothesis is that the entrepreneur is no longer a hero. He has become a socialized entrepreneur, an entrepreneur who is a player in an economy made up of on the one hand, the planning decisions made by big firms which, even if networked, nevertheless are powerful bureaucracies, and on the other hand, the state's policies to reduce unemployment and boost innovation and competitiveness.

Schumpeter, focusing on the entrepreneur considered as a *deus ex machina*, did not take enough account of the economic, political and social context which may give birth to the entrepreneurial function. He did not see how far the entrepreneur was a social agent and even sometimes a political construction, sometimes aiming at repairing the system's failures and extending its life. Today, and our approach aims at better explaining it, the innovative entrepreneur is replaced by a more pragmatic entrepreneur, whose main motivation lies in the acquisition of a revenue, which does not always mean his/her personal enrichment. Even if the conventional (in the sense of Galbraith) perception of the entrepreneur remains that of a hero, the one who can easily, thanks to his natural gifts, become a J.D. Rockefeller or Bill Gates, our approach is an attempt to demonstrate how this myth is an 'Innocent Fraud'.

NOTES

1. We use the male pronoun to avoid the clumsiness of 's/he' or 'he/she' every time, but we of course acknowledge that women entrepreneurs do exist.
2. 'The entrepreneur makes profitable the highest and humblest faculties of humanity. He receives the directions of the scientist and transmits them to the workman.' Author's translation.
3. J.A. Schumpeter defines the concept of 'creative destruction' which precisely highlights the capitalist craving for technology. 'This process of Creative Destruction is the essential fact about capitalism. It is what capitalism consists in and what every capitalist concern has got to live in' (Schumpeter 1942/1976, p. 83).

REFERENCES

Aghion, P. and P. Howitt (1998), *Endogenous Growth Theory*, Cambridge, MA: MIT Press.

Audretsch, D.B. (ed.) (2006), *Entrepreneurship, Innovation and Economic Growth*, Cheltenham, UK and Northampton, MA, USA: Edward Elgar.

Becker, G. (1975), *Human Capital*, Chicago, IL: University of Chicago Press.

Blaug, M. (1998), *Histoire de la pensée Economique*, Paris: Economica.

Bourdieu, P. (1980), 'Le capital social', *Actes de la Recherche en Sciences sociales*, **31**, 115–32.

Bourdieu, P. and L.J.D. Wacquant (1992), *An Invitation to Reflexive Sociology*, Chicago, IL: University of Chicago Press.

Boutillier, S. (2006a), 'Du capital social au potentiel de ressources, société entrepreneuriale, *quid* de l'économie sociale', *Cahiers du Lab. rii* no. 115, June, accessed January 2007 at http://riifr.univ-littoral.fr/wp-content/uploads/2006/12/doc%20115.pdf.

Boutillier, S. (2006), 'The end of capitalism: J.K. Galbraith vs K. Marx and J.A. Schumpeter', in B. Laperche, J. Galbraith and D. Uzunidis (eds), *Innovation, Evolution and Economic Change. New Ideas in the Tradition of J.K. Galbraith*, Cheltenham, UK and Northampton, MA, USA: Edward Elgar, pp. 53–71.

Boutillier, S. and D. Uzunidis (1995), *L'entrepreneur. Une analyse socio-économique*, Paris: Economica.

Boutillier, S. and D. Uzunidis (1999), *La légende de l'entrepreneur. Le capital social ou comment vient l'esprit d'entreprise*, Paris: Syros.

Boutillier, S. and D. Uzunidis (2006a), *L'aventure des entrepreneurs*, Paris: Studyrama.

Boutillier, S. and D. Uzunidis (2006b), 'Le potentiel de ressources de l'entrepreneur artisan dans un système territorial de production', *Cahiers du Lab. rii* no. 127, May, accessed May 2007 at http://riifr.univ-littoral.fr/wp-content/uploads/2006/12/doc127.pdf.

Boutillier, S. (2007), 'Marché et création d'entreprises. L'entrepreneur, "deus ex machina" du capitalisme?', in G. Rasselet (ed.), *Dynamique et transformations du capitalisme*, Paris: L'Harmattan, pp. 135–57.

Buchanan, J.M. and G. Tullock (1962), *The Calculus of Consent*, Ann Arbor, MI: University of Michigan Press.

Burt, R.S. (1995), 'Le capital social, les trous structuraux et l'entrepreneur', *Revue française de sociologie*, **XXXVL**, 599–628.

Canterbery, R. (2001), *A Brief History of Economics, Artful Approaches to Dismal Science*, Singapore: World Scientific Publishing Co.

Cantillon, R. (1997), *Essai sur la nature du commerce en général*, Paris: Institut National d'Etudes Démographiques.

Casson, M. (1990), *Entrepreneurship*, Aldershot, UK and Brookfield, US: Edward Elgar.

Castells, M. (2001), *La société en réseaux, L'ère de l'information*, Paris: Fayard.

Chabaud, D. and J. Ngijol (2005), 'La contribution de la théorie des réseaux sociaux à la reconnaissance des opportunités de marché', *Revue internationale des PME*, **18**, 73–90.

Coleman, J.S. (1988), 'Social capital in the creation of human capital', *American Journal of Sociology*, **94**, 95–120.

Coleman, J.S. (1990), *Foundations of Social Theory*, Cambridge, MA: Harvard University Press.

Davidsson, P. and B. Honig (2003), 'The role of social and human capital among nascent entrepreneurs', *Journal of Business Venturing*, **18**, 301–31.
Denison, E.F. (1962), *The Sources of Growth in the US*, New York: Committee for Economic Development.
Fukuyama, F. (1993), *The End of History and the Last Man*, New York: Harper Perennial.
Galbraith, J.K. (1967), *The New Industrial State*, Boston, MA: Houghton Mifflin.
Galbraith, J.K. (2004), *The Economics of Innocent Fraud*, Boston, MA: Houghton Mifflin.
Gasse, Y., M. Diochon and T. Menzies (2004), 'Les entrepreneurs émergents dont l'entreprise est devenue opérationnelle et les autres: comparaisons lors de la conception du projet d'entreprise', *Journal of Small Business & Entrepreneurship*, **17**, 117–34.
Granovetter, M. (1973), 'The strength of weak ties', *American Journal of Sociology*, **78**, 1360–80.
Granovetter, M. (1985), 'Economic action and social structure: the problem of embeddedness', *American Journal of Sociology*, **91**, 481–510.
Keynes, J.M. (1933), *The Means to Prosperity*, London: Macmillan.
Kirzner, I. (1973), *Discovery and the Capitalist Process*, Chicago, IL: University of Chicago Press.
Kirzner, I. (1985), 'Entrepreneurial discovery and the competitive market process: an Austrian approach', *Journal of Economic Literature*, **35**, 60–85.
Laperche, B. (2003) 'Risque et innovation: repères', in B. Laperche (ed.), *L'innovation orchestrée. Risque et organisation*, Paris: L'Harmattan, pp. 7–17.
Laperche, B. (ed.) (2005), *L'entreprise innovante et le marché. Lire Galbraith*, Paris: L'Harmattan.
Laperche, B. and D. Uzunidis (eds) (2005), *John Kenneth Galbraith and the Future of Economics*, London: Palgrave Macmillan.
Laperche, B., J. Galbraith and D. Uzunidis (eds) (2006), *Innovation, Evolution and Economic Change. New Ideas in the Tradition of Galbraith*, Cheltenham, UK and Northampton, MA, USA: Edward Elgar.
Lucas, R. (1988), 'On the mechanics of economic development', *Journal of Monetary Economics*, **72**, 3–42.
Macaulay, S. (1963), 'Non contractual relations in business: a preliminary study', *American Sociological Review*, **28**, 55–67.
Marchesnay, M. (2003), 'La petite entreprise: sortir de l'ignorance', *Revue française de gestion*, **144**, 107–18.
Marchesnay, M. (2004), 'L'hypofirme, fondement de l'hypermodernité?', 7e Congrès International francophone en entrepreneuriat PME, Paris, 27–29 October.
Marshall, A. (1890), *Principles of Economics*, London: Macmillan.
Marshall, A. (1906), *Principes d'économie politique*, Paris: V. Giard et T. Brière.
Marx, K. (1976), *Capital*, I, originally published in 1867, Paris: Editions Sociales.
Marx, K. (1976), *Capital*, III, originally published in 1894, Paris: Editions Sociales.
OECD (2001), *Du bien-être des nations, le rôle du capital humain et social*, Paris: OECD.
Porter, M. (1982), *Choix stratégiques et concurrence*, Paris: Economica.
Putman, R. (2000), *Bowling Alone: The Collapse and Revival of American Community*, New York: Simon and Schuster.
Romer, P. (1990), 'Endogenous technological change', *Journal of Political Economy*, **98**, 71–102.

Sammut S. (1998), *Jeune entreprise. La phase cruciale du démarrage*, Paris: L'Harmattan.
Say, J.-B. (1999), *Cours d'économie politique*, first published in 1820, Paris: GF-Flammarion.
Schulz, T.W. (1959), 'Investment in man: an economist's view', *Social Service Review*, **XXXIII**, 109–17.
Schumpeter, J.A. (1939), *Business Cycles*, New York: McGraw Hill.
Schumpeter, J.A. (1975), *Capitalism, Socialism and Democracy*, originally published in 1942, New York: Harper Perennial.
Schumpeter, J.A. (1983), *Histoire de l'analyse économique*, Tome II, originally published in 1954, Paris: Gallimard.
Schumpeter, J.A. (2006), *The Theory of Economic Development*, originally published in 1911, New Brunswick, NJ: Transaction Publishers.
Shane, S. (2003), *A General Theory of Entrepreneurship. The Individual-opportunity Nexus*, Cheltenham, UK and Northampton, MA, USA: Edward Elgar.
Smith, A. (2002), *Recherche sur la nature et les causes de la richesse des nations*, originally published in 1776, Paris: G-F Flammarion.
Suire, R. (2004), 'Des réseaux de l'entrepreneur au ressort créatif: quelles stratégies pour les territoires', *Revue internationale des PME*, **17**, 29–54.
Ulgen, F. (1996), 'Economie monétaire et dynamique schumpétérienne d'évolution économique', *Innovations, Cahiers d'économie de l'innovation*, **4**, 61–93.
Uzunidis, D. (1996), 'Schumpeter et mouvements longs de l'économie. Aspects critiques de l'actualité économique', *Innovations, Cahiers d'économie de l'Innovation*, **4**, 119–40.
Uzunidis, D. (2004), *L'innovation et l'économie contemporaine, Espaces cognitifs et territoriaux*, Brussels: De Boeck.
Uzunidis, D. (2006a), *Pépinière d'entrepreneurs: le capitalisme*, Paris: L'Harmattan.
Uzunidis, D. (2006b), 'Innovation, cycles et renouveau du capitalisme', in D. Uzunidis (ed.), *Pépinière d'entrepreneurs: le capitalisme*, Paris, L'Harmattan, pp. 71–87.
Walras, L. (1988), *Eléments d'économie politique pure*, originally published in 1874, Paris: Economica.
Weber, M. (1964), *L'éthique protestante et l'esprit du capitalisme*, originally published in 1904 and 1905, Paris: Plon.
World Bank (2000), *Attacking Poverty. World Development Reports 2000/01*, Washington, DC: World Bank.

10. Cooperative networks and clustering of high-technology SMEs: the case of Brisbane Technology Park

Kavoos Mohannak and Robyn Keast

1. INTRODUCTION

In today's global economy, innovation is presented as the key to economic development and sustainability. Only through innovation can companies ensure their progress and their place in the market and become strong enough to survive and prosper. By creating a favourable environment for innovation, countries and regions can facilitate their industries and companies to become stronger, be more profitable, and generate employment and sustainability.

The economic booms of some high-technology industrial agglomerations through the 1970s, both planned such as the Stanford Research Park and the Research Triangle Park and spontaneous such as Silicon Valley and the Cambridge Phenomenon, suggested a type of mechanism for creating a favourable environment for innovation. Aiming at recreating the dynamics found in those successful models in order to boost the economies of their perspective areas, more and more planned industrial agglomerations such as science and technology parks (STPs) have been established with strong government or quasi-government initiatives. These planned technology parks expected to provide the following benefits to the tenants: increased research productivity, employment growth in high-tech sectors, extraordinary growth or performance of R&D-intensive firms situated in the park, and the development of strong and operational ties between firms, university research, national laboratories and other research institutions. Many countries, including Australia, have invested in and developed technology parks with these objectives in mind. Some governments have also hoped that science parks will also help to: (a) raise the level of technological sophistication of local industries through the promotion of industrial R&D and promotion of foreign investments, and (b) accelerate the transition from a labour-intensive to a knowledge-intensive economy.

However, several recent studies have concluded that science parks tend to fail significantly in delivering the above widely expected benefits (Bakouros et al. 2002; Siegel et al. 2003; Hansson et al. 2005). If the role of science parks is not to contribute to creating growth and innovation by establishing and maintaining network relations between firms and academic institutions, what then if any is their role in a modern knowledge-based network society? It seems that science parks have generally failed to play a noteworthy function in establishing network relations between researchers and firms in the science parks.

In this regard this chapter aims to examine the nature and strength of network relationships of high-tech SMEs within an Australian technology park. By using a case study of an emerging south-east Queensland STP to provide insights, the chapter will focus on the nature of relationships and exchanges between SMEs and their stakeholders. Drawing on interviews, questionnaires, including a network linkage survey, as well as documentary material, this study further explores the role of cooperative networks and clustering in competency building and learning of high-tech SMEs in a planned cluster like the Brisbane Technology Park (BTP).

2. SCIENCE AND TECHNOLOGY PARKS: A BACKGROUND

A technology park is a term used to describe a variety of efforts to stimulate the development of entrepreneurial, knowledge-based SMEs. The main aim of a technology park is to help companies to become more innovative, therefore more competitive, and in so doing, increase the economic capacity of the region. Sanz (2002, p. 1) has noted the primary role of STPs is to 'actively support and facilitate the creation of new innovation-based local companies and to increase the competitiveness of the already existing local businesses'.

According to the United Kingdom Science Park Association's definition, a science park is a property-based activity configured around the following:

- formal operational links with a university or other higher educational or research institutions;
- the formation and growth of knowledge-based business and other organisations normally resident on site;
- a management function which is actively engaged in the transfer of technology and business skills to the organisations on site (UKSPA 1999).

Originally the terms technology or science park had a limited definition and referred specifically to the real-estate aspect of the park concept, in which universities typically leased real estate, office space and or research facilities to businesses. Over time the term has evolved to include a much broader range of functions, including economic development and technology transfer. However, in most cases the planned parks were established by artificial 'push' as the cause of regional technological development in the form of a type of infrastructure, usually to provide technical, logistic, administrative and financial assistance, which is envisaged as necessary for the establishment and competitiveness of high-tech or knowledge-based businesses.

It is true, however, that no two technology parks are alike. They differ in size, ownership, composition, location, services offered, types of restrictions used, appearance, relationship with universities, and other important characteristics. For example, Australian technology parks are small compared with overseas developments. In spite of this variety, the science and technology park movement shows the following trend during its 50-year development path:

1. Early science parks tended to be university initiatives located on university land. They first appeared in the United States for the purpose of better exploiting academic enterprise and research. Stanford Research Park, the first of the specialised science parks in the world, was established by Stanford University in 1951 (OECD 1987). The Research Triangle, a well-quoted case, was the creation of three universities already in that part of North Carolina with the state government (ibid.). A survey during the late 1980s found 61 out of 72 American research parks at that time to be either directly owned or affiliated with a university (Luger and Goldstein 1991). Pioneering parks in the UK, the Cambridge and Heriot-Watt Parks, were also started by their host universities in the early 1970s (Lowe 1985).
2. Later parks were characterised by greater involvement by local authorities concerned with restructuring the local economy and creating jobs. This type of park usually has a strong economic development objective. For example, many UK parks which were supported by public sector bodies are located in areas where voluntary private sector involvement is unlikely to be forthcoming (Grayson 1993). The potential for locational advantage through the attraction of high-technology firms to a disadvantaged area was a further incentive for such initiatives.
3. Following international trends in the 1980s, Australian state and territory governments developed explicit policy agendas to use science and technology parks as a tool to diversify the local economy, derive economic benefits and create new jobs from new firm creation and

new industries, and develop high-tech industry. As a result, many Australian parks such as the Technology Park in Western Australia, the Brisbane Technology Park (BTP), Tasmanian Technopark, and the Australian Technology Park (ATP) were sponsored by respective governments. The intention was to stimulate the development of high technology business by facilitating new starts in a designated locality – which may or may not be university spin-offs – and by encouraging existing high-tech companies to settle within these domains.

4. In recent years, increasingly the private property sector (both developers and investors) appears to be working in partnership with government, universities and other research centres in the development, procurement and property management of STPs (Zhang 2005). In addition new STP developments are being increasingly financed by private capital, sometimes in the form of loans or of direct investment. With fewer companies willing to enter into long-term lease commitments, it is likely that such property investors will become increasingly linked to the STP model.
5. There is also growing movement to drive public/private sector economic development in the direction of creating sustainable clusters. The cluster movement, with substantial public sector funding behind it, may provide a catalyst for the development of STPs in future (Heeres et al. 2004; Schoales 2006).
6. The newer parks also seem to involve a more proactive management approach to meet the changing need of potential tenants by providing more facilities such as multi-use incubator buildings for the growth of small firms and services such as technology transfer and funding advice. In this way, schemes developed in the late 1990s have come to be characterised by a more expanded, 'mixed use' model that includes academics, research, business, industry, commercial, government, and sometimes residential co-location within a park precinct (Giaoutzi and Vescoukis 2006).
7. STPs have also earned greater recognition in the 1990s from the collective activities of parks around the world. 'Sister parks' started to appear, which help develop corporate relationships, joint research and commercialisation on an international basis (Zhang 2005).

3. CONCEPTUAL FRAMEWORK

In this study the context and conceptual framework for the analysis of network relationships within a technology park is centred on the relevance of two key concepts for SME development: 'innovation clusters' and 'cooperation networks', recognising that both are emerging as significant tools to promote regional development through the activation, diffusion and expansion of locally generated knowledge. In the cluster/network-based approach, both concepts are linked by a focus on interactive learning and the diffusion of different types of knowledge: tacit or codified, scientific or practical etc. in different spatial and organisational settings. That also implies a focus on the emerging field of knowledge management that is the explicit and systematic management of knowledge and its associated processes of creation, organisation, diffusion and applications to create wealth and promote development.

The analysis is also guided by some basic assumptions: (1) SMEs can play a key role in triggering and sustaining economic growth and equitable development in developing regions; (2) the creation of technology-intensive firms is essential to build local capabilities to compete in the global economy and they are also essential to strengthen academic–industry–government linkages and encourage technological innovation; and (3) the region's development potential can be greatly enhanced by adopting a cluster/network-based approach to address its development needs and spatial imbalances, searching for cooperation and partnerships among different government levels, the private sector and international organisations (see Kaufmann and Todtling 2002).

SMEs, particularly technology-based ones, have great potential to accelerate economic growth and expand their share of exports. However, this potential role is often not fulfilled because of their small scale. As Ceglie and Dini (1999) point out, SMEs are often unable to capture market opportunities that require a large scale of production. They are also unable to achieve economies of scale in the purchase of inputs (such as equipment, raw materials, finance, consulting services, etc.), and the creation of an internal division of labour that could foster cumulative improvements in productive capabilities and innovation.

Small size also constrains the internationalisation of dynamic functions such as training, market intelligence, logistics and technology innovation. Even innovative technology-based firms tend to lack key skills and resources, such as marketing or business capabilities. Berry (1997) notes that small-firm limitations typically fall in the areas of: access to technological information; guidance on quality control; access to finance; assistance in purchase of materials or equipment; and market stability.

It is widely acknowledged that inter-firm cooperation and linkages involving SMEs may have a strong impact on growth and their performance (e.g. Longhi 1999). The development of networks can improve the competitive position of SMEs and reduce the problems related to their size through mutual help. For example, firms may establish a localised network to become more specialised and complement each other's capabilities by sharing resources, pooling together their production capacities and purchasing power, thus achieving scale economies to conquer markets beyond their individual reach (Pyke 1992).

3.1 The Role of Innovation Clusters

As argued widely in the literature, horizontal cooperation and the creation of external economies among SMEs in clusters contribute to generate competitive advantage through 'collective efficiency' (e.g. Schmitz 1995; Simmie and Sennett 1999). Schmitz (1995) emphasises that external economies are essential to growth but not sufficient to ride out major changes in product or factor markets; that would require joint action. The advantages of cooperation among SMEs are usually connected with collective economies of scale, the benefits of dissemination of information and inter-firm division of labour. These benefits tend to increase when transaction costs are low, and these in turn tend to decrease with geographic proximity and the establishment of shared infrastructure, common norms and tacit rules for cooperation.

Therefore, an innovation cluster such as a technology park is characterised not only by a sectoral and geographical concentration of firms and other economic agents which gives rise to external economies and favours the creation of specialised technical and financial services, it also comprises public and private local institutions to foster knowledge transfers and to support local economic development. This type of arrangement facilitates collective learning and innovation through implicit and explicit coordination. Successful clusters depend on both the private and the public sector (usually universities and research institutions), that join efforts to create innovative environments and build synergies among agents with complementary capabilities (Perrons 2004). Their development is gradual and cumulative; over time the region builds knowledge, skills, institutional support structures, specialised services, financing arrangements, infrastructure and collective norms of cooperation and mutual trust. This cooperative rather than wholly competitive orientation allows cluster members to leverage the expanded knowledge set and resource base made available through their relationships in order to create innovative results.

In this way, clusters are built on linkages and relationships that integrate the isolated technological capabilities of institutions, firms and individuals into a collective spatial asset. The establishment of mechanisms for effectively and efficiently coordinating these relationships is essential to create a supportive environment for many forms of technical interchange, cross-fertilisation, risk-sharing and collective learning. This is essentially a territorially based process, as people who share the same space discover the advantages of 'learning by interacting' and synthesising by sharing (Steinsli and Spilling 2004).

As Rabellotti (1995) points out, the crucial characteristic of a cluster or milieu is the set of competitive and collaborative linkages among agents in a socially and historically defined agglomeration, complemented by a set of collective intangible assets that belong to the production system as a whole. The cluster benefits from its complex web of interactions because innovation rarely happens in isolation.

Among the diverse 'locational ingredients' that are usually associated with high-tech cluster and technology park development, two factors may be considered as necessary but not sufficient conditions for a successful cluster. The first is a 'critical mass' of human resources, including entrepreneurs, scientists, engineers, technicians and skilled labour. The second is a capable scientific and technological infrastructure, or the 'knowledge assets' of a region. These may include universities, public and private research labs, technological incubators and innovation centres. The main roles of these anchor institutions are to promote technology transfers and to support networking.

4. METHODOLOGY

The Brisbane Technology Park (BTP) has been purposely chosen to form the basis of this study for a number of reasons. First, although the BTP has been in existence since the 1980s, it has experienced a number of fluctuations in terms of its membership composition and related cooperative relationships. It is envisaged that this variation in membership and cohesion will provide useful insights into processes for building and maintaining cooperative networks. Second, as BTP is not located in direct proximity to universities and other such research institutions, valuable network connections to such institutions will require purposeful establishment by BTP members. Furthermore, BTP is an initiative of the 'Department of State Development, Trade and Innovation', and forms part of Queensland's 'Smart State' programme, designed to create and stimulate growth in employment and the state economy. The key objectives of BTP as outlined on its official website are:

- Develop as a community that makes visible and evident a range of business, economic and scientific skills;
- Provide resident companies involved in the commercialisation and exploitation of technology with a sense of identity; and

- Provide an environment that develops a sense of community in which all residents contribute to the healthy exchange of scientific and other knowledge between businesses and individuals (BTP 2006).

Such objectives appear consistent with those of the majority of technology parks identified within the literature, in that most are seeking to achieve economic returns at the park and regional level (Phillips 2002). Another feature common to STPs is the presence of a park management body, charged with the task of maintaining infrastructure and driving the region to meet set objectives; in the case of BTP, a combination of a public and a private sector organisation is involved in both park precinct management and park development management, to ensure both state government and park objectives are achieved.

For the review, a combination of quantitative and qualitative data was collected, by means of a questionnaire and face-to-face interviews. The structured questionnaire was designed to identify the nature and type of relationships of BTP firms, the perceived importance of these linkages, and to map the scope of cooperative networks. The questions were formulated in two different ways: (a) 'factual' questions requiring a dichotomous (yes/no) response; and (b) questions which were answered on scale-type responses, indicating intensity of the linkages, location and importance of the information and knowledge ranging from little importance to very important. Additionally, follow-up discussions and interviews held with senior managers of selected firms were undertaken in order to investigate in more qualitative detail the nature of their relationships and their experience of the learning and sharing processes within networks.

The sample of firms used in this study was drawn from the directory of companies available on the BTP's website. The firms within BTP largely belong to the sectors of multimedia, software development, electronic services, telecommunications and biotechnology; i.e. so-called high-tech firms. All 56 of the firms located with BTP at the time of this study were chosen to receive questionnaires. The five-page questionnaire was first emailed to companies in BTP. Follow-up telephone calls and emails were also made. In total, 24 usable firm responses were received, yielding a response rate of 43 per cent. Out of these firms 13 are regular SMEs, while 10 are branch offices of existing firms and only one is a corporate spin-off. Among these firms 10 have less than 20 employees, nine between 20 and 100, and four have over 100. The relatively low response rate can be explained in part by a period of turbulence within BTP, during which several members entered and exited the cluster, making it difficult to gain access to firms active within the park.

5. NATURE OF NETWORK LINKAGES

In order to achieve a detailed understanding of the type and nature of the relationships occurring between BTP members and their impact on the exchange of knowledge and the fostering of innovation within the park, a set of questions based on network linkages and relationships was posed. The first question, administered in the form of a network linkage survey contained within the questionnaire, provided a list of possible BTP members and asked respondents to indicate with which organisations they had regular contact/exchanges against a set of variables such as type of interactions (e.g. technical advice, informal contacts, etc.) (see Table 10.1).

As Table 10.1 displays, BTP firms are involved in an array of linkages, of which exchange of information (51), informal contacts (42) and formal contracting (42) are the most frequently reported. This result is not surprising since shared information exchange is understood to be the most common and least risky form of exchange between organisations, and as Pake (1986, p. 36) has noted, 'Beneath most formal ties . . . lies a sea of informal relations'. It should be noted, however, that while at first glance it appears that participating firms are involved in a high level of interaction; on average it is quite low in comparison to other studies (Schmitz 1995). For example, each of the 24 firms indicated that they engaged in formal contracts with only 2.33 other firms of the total of 56 located in the BTP at the time of the study. Nevertheless, the existence of multiple ties is an indication of a level of strength of ties occurring between those firms, providing a stronger basis for cohesive activity toward innovation outcomes (Powell et al. 1996). Further, a more detailed interrogation of the linkages uncovers the existence of a core group that is involved in multiple types of exchanges. This result provides some evidence that several park members have entered into relationships based on higher levels of trust, involving the exchange of information and technological advice, suggesting that some firms have been able to achieve strong ties within the BTP.

Table 10.1 Aggregate linkages between BTP members

Type of interaction	Aggregate number of interactions among the whole sample (N=24)
Technical advice	16
Informal contacts	42
Exchange of information	51
Formal contract	42
Joint funding	6

Table 10.2 Linkages with other networks

Networks	High	Medium	Low
BTP network/cluster	3	10	9
Local network (Brisbane and surrounds)	11	6	6
National	13	8	3
Global	14	4	5

Additionally, participants were asked to identify any major links with other outside networks and rate the importance of these in terms of their business activity; three rating levels, high, medium and low, were provided. It is clear from the data collected that networks for BTP firms extend beyond the local region (see Table 10.2).

Nearly 50 per cent of the firms reported the importance of national and global networks. Data suggest that most of these networks are between firms in the supply chain (suppliers and customers). This implies that, despite acknowledging the value of local 'networking' between firms in the park, firms have not actively involved themselves in localised networking interaction as a particularly important factor in their general business activities. This emphasis on external relations was underscored in interviews carried out with firms within the park. For example one respondent noted:

> Suppliers, most of them are international, ranging from Italy to Canada to the United States and the United Kingdom. Then my clients are basically Australian and New Zealand. (interview, 24 May 2005)

As well as providing specific support services and resource advantages, linkage to additional external networks can facilitate access to knowledge and opportunities. Because regional and broader institutions interact within relatively larger networks of linkages, they can act as innovation facilitators. Evidence of the utilisation of external networks to obtain knowledge has been identified in interviews conducted in this study, with one firm suggesting they had easier access to knowledge other than within the technology park:

> Inwardly we have a lot of knowledge within the company, so there are very few occasions where it would require me to go to anyone within the Park, not because they haven't got the knowledge, don't get me wrong about that. It's just that I have easier access to knowledge in other places. (interview, 13 April 2005)

While this information does suggest some element of strong ties may exist between some BTP firms, in general, relationships appear to be established and governed principally by economic transaction contracts, highlighting the tension between competition and cooperation within such entities. Further, the findings broach the fact that BTP firms appear more interested in building on relationships outside the park than tapping into the benefits available locally; this phenomenon could be referred to as an 'innovation slippage' in which potential advantages to cluster members have slipped out of the local network and entered the broader/distant network. This notion of 'innovation slippage' within the BTP may suggest that a lack of commitment to valuable knowledge exchange exists within the park.

To analyse the degree of willingness and commitment to exchange knowledge and achieve economic advance within the BTP, respondents also were asked if they believed that friendly contacts with other firms in the BTP were an important asset for their firm; approximately 65 per cent of respondents indicated a perceived value in their relationships within the BTP. Some uncertainty as to the source of this value, however, was evident, as indicated in the following interview extract:

> It's difficult to say just what the advantage will be but as I just said there are advantages. To discover that someone's got some products that you can use, or even that you can manufacture the product that the other guy wants to do, we've got software here, you've got developers, we've got testers and graphic parts, if a small company grabs that, there has got to be advantages. (interview, 23 May 2005)

In this way, there is an evident recognition of the potential benefits of relationships within the BTP, particularly given knowledge of other people and their products; however, due to the uncertainty of the advantages to be derived from relationships within the park, trust (or a 'leap of faith') is required to take steps to obtain these advantages.

Interestingly, when asked if they felt they could achieve the same outcomes if located outside of the park (16 respondents indicated they had developed new products and processes while located at the BTP), 68 per cent of respondents believed they could. This finding is particularly noteworthy given the emphasis in the literature on the difficulties faced by SMEs operating autonomously; one explanation of this finding is that several of the interviewed firms were branches of existing firms, which may be capable of providing them with the required support for innovation. Finally, respondents were asked to detail the frequency with which they exchanged ideas or experiences with other firms – a majority of them (16 firms) responded only 'occasionally' and only three firms indicated 'often'. The responses of six firms, however, indicated that they have never exchanged ideas or experiences with other firms.

In summary, this case study has presented a range of findings on the nature and type of networked relationships within technology parks, and in particular the BTP. While the respondents indicated that they valued the provision of supportive infrastructure (consistent with the literature previously highlighted), there was some deviation from the literature in a number of areas, principally relating to the amount of emphasis placed on external networking rather than developing internalised cooperative networks (Boland and Tenkasi 1995, pp. 354–5), sometimes referred to as 'communities of knowing'. As a consequence, BTP firms may be failing to capitalise on the opportunities presented to them locally, inevitably also costing them greater amounts of time and money necessary to develop geographically dispersed relationships.

6. CLUSTER GOVERNANCE AND STRENGTH OF RELATIONSHIPS

As well as stressing the importance of a cooperative or collective orientation, the literature also indicates that the structural aspects and operational characteristics of a cluster type of arrangement such as an STP are essential for collective learning. In this regard, the optimal structure combines both stability and variety. Arriving at such an optimal structural mix generally requires oversight and direction setting of a governance model or framework. There is an array of governance structures that can be used to organise the transactions between business entities in a cluster. In addition to the three major mechanisms of vertical integration and legal authority, the contractual basis of the market and the interpersonal characteristics of networks, hybrid arrangements that draw on aspects of two or more modes have come to the fore (Powell 1990). The choice of governance mode within a cluster has important consequences for both facilitating knowledge sharing and learning and its efficient and effective operation and management.

In order to gain insights into the type of the mechanisms and structures that may provide effective integration strategies respondents were asked to rate the importance of possible mechanisms for the successful integration of a cluster like the BTP. Specifically the respondents were asked to indicate those linking structures and/or processes that may assist the way the BTP network functions. A list of 11 possible answers was provided, with respondents asked to indicate those most relevant to their situation within the BTP. The replies to the questions are set out in Table 10.3.

These findings, when coupled with other results, indicate that although there is evidence of a market orientation based on exchanges, and some centralised authority by way of the BTP government sponsorship and

Table 10.3 Importance of linkage mechanisms

Linking mechanisms	Importance (N=24)*
Well developed rules and procedures	4
Ongoing liaison between members' agencies	**9**
Coordination committees	7
Involvement of a facilitator (intermediary)	4
Involvement in working parties	4
Conferences to discuss issues and set joint direction	7
Workshops to develop mutual implementation strategies	6
Informal meetings between members	**13**
Participation in joint venture programmes	6
Regular telephone contact	4
Regular meetings of cluster members	**8**

Note: * Multiple choices allowed.

management direction, the BTP is also guided by a relational-based governance mode. This result is consistent with the literature that stresses the importance of a relational element to network/cluster governance (Jarillo 1988; Lowndes and Skelcher 1998).

Further highlighting the network orientation within the cluster, the responses have also demonstrated a preference by respondents to utilise highly personalised linkage mediums, for example ongoing liaison (9) and informal meetings (13), to provide the linkage mechanism to bind members of the BTP. These were bolstered by more formalised but nevertheless interpersonally orientated mechanisms such as regular meetings, conference and workshop participations. One of the interviewees indicated that:

> When I first came that was my idea to get some networking going, but being focused on getting going I sort of dropped the ball in that respect and started to concentrate on my own business. But I do believe that somebody perhaps needs to provide the wherewithal to get people invited to a morning coffee where we could discuss common problems and from there grows the faculty exchange – 'Oh, you can do that and I can do this'! (interview, 23 May 2005)

This finding reflects the importance of relationships as a mechanism for development and points to the need for strategic attention directed toward building and leveraging relationships within the cluster to maximise information exchange and knowledge synthesis. The ability to manage and drive relationships toward a collective outcome is confounded by the often oscillating orientations and commitment levels of cluster members.

6.1 Strength of Relationships

Based on the data collected during the study of BTP, it is proposed that many of the members of this particular technology park may have entered for reasons other than just the collaborative advantage of cluster locales. For example, the interviews revealed that an attractive location and tenancy rental costs, as well as the association with the positive image of a technology park, were considerations for residency within the BTP (interviews, 13 April 2005).

While technology parks offer members a wide range of benefits, including those identified above, their propensity to facilitate networking amongst members is generally seen as the greatest advantage. Co-location within STPs additionally promotes the transference of more complex tacit knowledge, which may lead to opportunities for innovation, one of the primary goals of a majority of technology parks, including the BTP. In the case of a number of surveyed and interviewed BTP members, however, their focus, at least initially, was on the establishment of their own businesses, rather than on networking (that is, establishing contacts and learning about the operations of others and their resources) and searching for potential synergies within the park.

As well as the rationale for agglomerating within a locale, the nature and strength of the relations between clustering organisations has been shown to have an effect on securing innovative outcomes. Theoretically speaking, Granovetter (1973) and Uzzi (1997) outline some characteristics of ties. According to Granovetter (1973, p. 1361): 'the strength of a tie is a (probably linear) combination of the amount of time, the emotional intensity, the intimacy (mutual confiding), and the reciprocal services which characterize the tie', while Uzzi (1997) identifies levels of trust, fine-grained information transfer, and joint problem-solving arrangements as important criteria of the strength of ties between firms. But, in the end, as Granovetter (1973) contends: 'it is sufficient . . . if most of us can agree, on a rough intuitive basis, whether a given tie is strong, weak, or absent.' According to the weak-tie concept originally advanced by Granovetter, distant and infrequent relationships (i.e. weak ties) are efficient for knowledge sharing because they provide access to novel information by bringing otherwise disconnected groups and individuals. Strong ties, by contrast, are likely to lead to redundant information because they tend to occur among a small group of actors in which everyone knows what the others know.

Based on these criteria, and drawing on the relational data highlighted in Tables 10.1 and 10.2, it can be seen that the BPT exhibits relatively weaker ties or relationships within its local domain and stronger relationships with external bodies. To expand, although ranging across a variety of tie types, within the cluster locale there was an average of only 2.33 interactions, between member bodies, suggesting a weak relational bond. By contrast, as Table 10.2 depicts, there appears to be a stronger relational attachment by BTP members to other organisations and networks external to the park precinct. In this way, it could be argued that the BTP with its strong ties to external and weaker internal ties, does not present an ideal or efficient cluster structural composition.

However as has been identified in the literature, firms in STPs need a variety of network ties ranging from loose/weak ties designed to identify new opportunities, through to strong ties based on trust and cooperation designed to facilitate the innovation process (Koh et al. 2005). Government policies designed to improve access to technological knowledge often encourage firms to develop strong ties with competitors, suppliers or customers. But although strong ties are valuable, especially when tacit knowledge needs to be communicated, firms may resist entering into close relationships with other firms, because of the potential for exploitation. As a result, policies that encourage such firms to form weak ties may be a more effective way of promoting the spread of technological knowledge. Furthermore, these weak ties can act as a precursor to expanding network strength by helping cluster members to establish initial 'low risk' linkage relationships which can subsequently be built upon to create the bonding environment necessary for innovation development.

The important question then is what kinds of ties need to be established for various network arrangements, especially in cluster, as in a technology park. In fact, a great deal of recent government policy has been devoted to encouraging firms to build strong ties. In many cases, this has been aimed at establishing intra-industry networks that would allow firms to diffuse technological knowledge more widely than would occur of there were only weak ties, or no ties at all. As Uzzi (1997) notes, strong ties offer great advantages for the spread of knowledge. They encourage joint problem-solving activities that combine diverse talents with scarce technological knowledge to broaden the range and improve the quality of R&D activities. Second, because of the possibility of achieving exchanges of knowledge and information that are more fine-grained, strong ties can overcome the problems associated with tacitness. Finally, strong ties are associated with higher levels of trust, which can make the parties more willing to exchange valuable knowledge than they would in arm's-length relationships.

However, despite the importance of the strong ties, it is argued that weak ties can also be of value to all firms and the encouragement of weak ties in general may prove especially beneficial to firms whose management has, for example, an individualistic style. Weak ties also can benefit originators of new technological knowledge by bringing them into closer contact with a wider selection of possible users (Robertson 1998). The task then, for those responsible for the design and administration of clusters, is to enable the correct relational fit between weak and strong ties necessary for optimal innovation development and diffusion.

6.2 Building Networks: Policy Implications

As discussed, a major factor of any high performance cluster/park is the existence of extensive informal and formal relations between firms and even competing firms. This extends across the cluster, between firms and their supporting infrastructure.

Soft networks (such as local professional and trade associations) and hard networks (strategic alliances between firms) are both important and the development of such networks is supported by local culture that enables both competition and cooperation and even collaboration to thrive. In fact, the balance between competition and cooperation within clusters (i.e. coopetition) is an important determinant of the direction of policy. In countries like Australia, with more flexible labour markets and 'orthodox' competitive norms, collaborative arrangements between firms may be less likely to develop. Some mechanisms of fostering cooperation, including most of those available to regional and local policy-makers, focus upon the support of public–private collaborations and the promotion of collective learning processes. At one level, this can be regarded as requiring the encouragement of networks and partnerships between large companies, the SME sector, trade associations, universities and research institutes.

The establishment of networks enables firms in each cluster to expand their access to knowledge and resources from a broader range of sources, in addition to expansion of the resources available to the parks and clusters themselves. It becomes clear that a broad-based cluster/network-based system may have a strong impact in specific areas such as the following, which was identified as critical in this study:

- Access to new markets and marketing strategies;
- Access to capital: integrated access to services such as financial planning, support for obtaining grants, opportunities for access to venture, development, and seed capital;
- Expansion of inter-firm linkages: a networked approach is ideal for maximising the impact of programmes and projects, such as partnerships, alliances, and linkages to outside suppliers;
- Technological support: access to services such as technology assessment and forecasting, assistance on technological choices, marketing

assessment of innovative projects and access to outside technical information;
- Technology transfer opportunities: networks may be used to stimulate investment in science and technology, R&D, technology transfer and spin-offs;
- Access to talent and know-how: networks may help in the process of identifying and hiring skilled people across regional boundaries;
- Strengthening local cluster governance structures: the establishment of linkages with other clusters/parks would enable a better understanding of stakeholder needs and markets and may be used to disseminate best practices in technology parks to improve the performance of firms in each cluster/park;
- Optimising and sharing facilities: the operational support infrastructure may be optimised and many facilities could be shared over the network, including incubators, prototype centres, pilot plants, test laboratories, and online conferencing facilities.

Some of these benefits have been achieved in the case of the Brisbane Technology Park. For example, the establishment of localised technology support mechanisms, such as the Australian Institute for Commercialisation (AIC), build on regional strengths to stimulate innovation activities and diffuse technology to SMEs. However, local agents should now expand cooperative agreements, alliances and consortia involving public institutions, local firms and foreign organisations. It is clear that firms now recognise the need to develop and strengthen technology information networks, technology transfer mechanisms and liaison agents to promote exchanges of knowledge among firms, research institutions and regions. Clustering is not just strategic plans or meetings, but more important, a means of quickly promoting knowledge and competence flows. The lesson for government and other bodies responsible for the administration and architecture of clusters appears to be that these entities cannot be manufactured – that is simply providing real estate and a broad policy direction. However, conditions can be established that facilitate the formation, management and sustainability of clusters, accelerate knowledge sharing and innovation development and maximise their contribution to economic/regional development.

7. DISCUSSION AND CONCLUSIONS

In the context of a systematic innovation approach this study has investigated cooperative networks of a high-tech SME-based innovation cluster – the Brisbane Technology Park. In meeting the objectives of this study this chapter has explored the benefits of cooperative networks and clustering. In terms of the existence of cooperative networks that should contribute to the competencies of high-tech SMEs, the evidence is quite ambiguous. For example, co-location does not, in any straightforward way, appear to translate into cooperative network endeavours, although clearly there is some evidence of such outcomes. This suggests that these firms do not appear to operate in the typical innovation cluster way suggested by the literature. It seems that for these firms, the existence of skilled employees, in-house R&D and finance, are more crucial factors encouraging co-location or agglomeration. As the respondent firms indicated, the existence of localised human resources with strong technological capability is of great importance for their innovative activities.

As for the importance of locality, it appears that the regional range of linkages is focused primarily on national and global levels. While linkages with customers and suppliers are very intensive with partners across Australia and overseas, collaboration with universities and training institutions often takes place in the closer vicinity. This is probably due to the tacit character of knowledge transfer between these partners, where face-to-face contacts are a prerequisite for joint innovation projects. These findings suggest that while these firms operate within a local cluster, substantial links exist to suppliers and customers outside the region. This is particularly true for more innovative firms that seek to extend their operation nationally and globally.

Ensuring effective technological connectivity coupled with appropriate management know-how is essential for these high-tech firms. It may enable them to connect more adequately, particularly in their knowledge base, learning processes and competencies, with their business environment. Also, from a policy development point of view it is critical to understand the evolutionary stages of innovation in the firm as well as the cluster in which the firm is located. Of equal value is the recognition of each stage in the evolutionary process that offers its own opportunities for managing innovation both at the level of the firm and at the cluster level.

In sum, by studying the examples of clustering in a technology park, both regional policy-makers and firm managers can make decisions aimed at improving the innovation process and competitiveness capabilities. In forming effective policies at the level of both the firm and the region, decision-makers may need to consider the learning process inherent in the management of externalities and internalities, especially in dealing with uncertainty and complexity.

Finally, it is important to consider the main critical factors of innovation success in the context of institutional linkages and cluster/technology park theory. In fact, although knowledge clusters are highly individual and differentiated, it is helpful to identify some common factors of success and, in particular, their innovative capability, linked to continuous learning at the level of the single firm and of systems of firms, which refer to collective learning and learning by interacting. Some of these factors are mentioned below. These factors can be grouped as 'preferred location factors', 'strategic micro-management factors' and 'supportive macro-environment':

Preferred location factors This set of factors generally influences selection of business location for knowledge-based firms. As mentioned before they include availability of advanced IT infrastructure, R&D facilities, skilled and specialised labour forces. Some of the factors also relate to the quality of the work inside and outside the park such as access to the airport and business image.

Strategic micro-management factors These relate to the professional management of the park and also the need to have a clear park objective. For this purpose, it is important to have a tenant-oriented management style. A park management team has to be aware of their tenants' needs so as to assist them when required. Besides, it is necessary to have a clear target about park achievements such as number of R&D successes, tenant growth, contribution to regional economy, or number of technology transfers.

Supportive macro-environment Cooperative networks and support from various related and supporting industries and sectors are important for a technology park to achieve success. Linkages and cooperation can enhance firms' innovative capacity and effectiveness, and lead to their long-term competitiveness. Government is also a key player in constructing such a supportive environment. As emphasised in the literature and through interviews with tenants it is especially decisive for the government to provide support in such aspects as securing an attractive park site; providing financial support not only for park establishment and development but also for channelling firms to the park through incentives and support, and providing tenants with access to legal and technological support services. Government can also call on related and supporting industries to provide cooperation and support for knowledge-based firms within the park.

The study has shown that technology parks in general and the BTP in particular should expand their outreach role. Successful technology parks need to be integrated into local infrastructure but also to national and global sources of technologies and markets. Furthermore, the evidence strongly suggests that all forms of networking should be fostered at diverse spatial levels. This may range from local venture forums to bring together potential investors and local firms to links with other regions as a way to broaden their sources of information and to develop new markets.

Even though the impacts of the programme are just beginning to appear, this case illustrates how governments can contribute to the development of adaptable programmes and processes, by acting as both enablers and facilitators, and brokering new kinds of networking and association. For that purpose, clusters and technology parks may become not only significant tools to foster SME competitiveness, but also to help establish a new model of regional development, based on the positive impacts of sharing knowledge, advanced skills and innovation. It is only through cooperative networks that innovation clusters can leverage collective resources and secure competitive advantage.

REFERENCES

Bakouros, Y.I., D.C. Mardas and N.C. Varsakelis (2002), 'Science park, a high tech fantasy? An analysis of the science parks of Greece', *Technovation*, **27**, 123–8.

Berry, A. (1997), 'SME competitiveness: the power of networking and subcontracting', Inter-American Development Bank occasional paper no. IFM-105, Washington DC.

Boland, R.J. and R.V. Tenkasi (1995), 'Perspective making and perspective: taking in communities of knowing', *Organization Science*, **6**, 350–72.

Brisbane Technology Park (BTP) (2006), 'Brisbane Technology Park – about the park', accessed 15 January at www.brisbanetechnologypark.com.au/about.asp.

Ceglie, G. and M. Dini (1999), 'SME cluster and network development in developing countries: the experience of UNIDO', presentation to International Conference on Building a Modern and Effective Development Service Industry for Small Enterprises, 25 March, Rio de Janeiro.

Giaoutzi, M. and V. Vescoukis (2006), 'The role of ICTs in building entrepreneurial capacity in the regions of the future', *International Journal of Entrepreneurship and Innovation Management*, **6** (3), 191–204.

Granovetter, M.S. (1973), 'The strength of weak ties', *American Journal of Sociology*, **78**, 1360–80.

Grayson, L. (1993), *Science Parks: An Experiment in High Technology Transfer*, London: British Library Publications.

Hansson, F., K. Husted and J. Vestergaard (2005), 'Second generation science parks: from structural holes jockeys to social capital catalysts of the knowledge society', *Technovation*, **25**, 1039–49.

Heeres, R.R., W.J.V. Vermeulen and F.B. de Walle (2004), 'Eco-industrial park initiatives in the USA and the Netherlands: first lessons', *Journal of Cleaner Production*, **12**, 985–95.

Jarillo, J.C. (1988), 'On strategic networks', *Strategic Management Journal*, **9** (1), 31–41.

Kaufmann, A. and F. Todtling (2002), 'How effective is innovation support for SMEs? An analysis of the region of Upper Austria', *Technovation*, **22** (3), 147–59.

Koh, F., T.H. Winston Koh and F.T. Tschang (2005), 'An analytical framework for science parks and technology districts with an application to Singapore', *Journal of Business Venturing*, **20** (2), 217–39.

Longhi, C. (1999), 'Networks, collective learning and technology development in innovative high technology regions: The case of Sophia-Antipolis', *Regional Studies*, **33** (4), 333–42.

Lowe, J. (1985), 'Science parks in the UK', *Lloyds Bank Review*, **156**, 31–42.

Lowndes, V. and C. Skelcher (1998), 'The dynamics of multi-organisational partnerships: an analysis of changing modes of governance', *Public Administration*, **76** (2), 313–33.

Luger, M. and H.A. Goldstein (1991), *Technology in the Garden, Research Parks and Regional Economic Development*, Chapel Hill, NC and London: University of North Carolina Press.

OECD (1987), 'Science parks and technology complexes in relation to regional development', general distribution report, Paris: OECD Publications.

Pake, G.E. (1986), 'Business payoff from basic science at Xerox', *Research Management*, **29**, 35–40.

Perrons, D. (2004), 'Understanding social and spatial divisions in the new economy: new media clusters and the digital divide', *Economic Geography*, **80** (1), 45–52.

Phillips, R. (2002), '400 tech parks help firms act globally', *Research-Technology Management*, **45**, 5–6.

Powell, W.A. (1990), 'Neither market, state or hierarchy: network forms of organization', *Research in Organizational Behaviour*, **12**, 295–336.

Powell, W. K. Koput and L. Smith-Doer (1996), 'Interorganisational collaboration', *Administrative Science Quarterly*, **41**, 116–45.

Pyke, F. (1992), *Industrial Development through Small-Firm Co-operation*, Geneva: International Institute for Labour Studies.

Rabellotti, R. (1995), 'Is there an industrial district model? Footwear districts in Italy and Mexico compared', *World Development*, **23**, 29–41.

Robertson, P.L. (1998), 'Information, similar and complementary assets, and innovation policy', in Foss N.J. and B.J. Loasby (eds), *Economic Organisations, Capabilities and Co-ordination: Essays in Honour of G.B. Richardson*, London: Routledge.

Sanz, L. (2002), 'The role of science and technology parks in economic development', e-report by International Association of Science Parks, accessed 15 January, 2006, at www.iasp.ws.

Schmitz, H. (1995), 'Collective efficiency: growth path for small-scale industry', *Journal of Development Studies*, **31**, 529–66.

Schoales, J. (2006), 'Alpha clusters: creative innovation in local economies', *Economic Development Quarterly*, **20** (2), 162–77.

Siegel, D.S., P. Westhead and M. Wright (2003), 'Assessing the impact of university science parks on research productivity: exploratory firm-level evidence from the United Kingdom', *International Journal of Industrial Organization*, **20**, 177–84.

Simmie, J. and J. Sennett (1999), 'Innovative clusters: global or local linkages?', *National Institute Economic Review*, **170**, 87–98.

Steinsli, J. and O.R. Spilling (2004), 'On the role of small firms in cluster evolution: the case of Internet development in Norway during the 1990s', *International Journal of Entrepreneurship and Innovation Management*, **4**, 194–215.

UKSPA (1999), '15th Anniversary 1984–1999', report for the United Kingdom Science Park Association, Birmingham.

Uzzi, B. (1997), 'Towards a network perspective on organizational decline', *International Journal of Sociology and Social Policy*, **17**, 111–55.

Zhang, Y. (2005), 'The science park phenomenon: development, evolution and typology', *International Journal of Entrepreneurship and Innovation Management*, **5**, 138–54.

11. Clumps or clusters: a case study of biotechnology and life sciences in the Seattle area

Paul Sommers

1. INTRODUCTION: CLUMPS AND CLUSTERS

Bob Drewel, the leader of a regional economic development organization in the Puget Sound, says that he is trying to encourage clusters of firms in the region, not clumps of firms.[1] 'Clumps' in his view are firms that just happen to be located near each other but do not interact much to exploit their potential agglomeration economies, whereas clusters deliberately exploit a variety of interdependencies that make themselves and the region more competitive. This chapter examines this distinction, using the case of biotechnology-related firms and institutions in the Seattle area as a test of the theoretical perspective developed below.

Drewel identified one of the key issues in the debates around cluster initiatives. Michael Porter's classic cluster model focuses attention on competitive relations among firms in an industry striving to meet the needs of demanding customers, and among suppliers to those firms, all of which are located in proximity to each other. He also stresses the importance of supporting institutions in his well-known diamond model (Porter 1990, 2000). Stuart Rosenfeld, on the other hand, has emphasized the importance of collaboration among firms and key institutions in a regional cluster:

> The glue that continues to make proximity matter consists of 'soft' externalities, e.g. greater access to tacit knowledge, opportunities for deliberate acts of cooperation and collaboration that give companies the strength of numbers to influence customers, markets, or policies, and access to experienced labour. (Rosenfeld 2005, p. 6)

Both of these theoretical frameworks imply that firms and institutions within a cluster are aware of their interdependencies and use either competitive or collaborative mechanisms to gain benefit from these interdependencies.

In contrast, many cluster studies rely on empirical rubrics for identifying potential interdependencies, rubrics that do not directly test for awareness of interdependencies on the part of cluster members nor exploitation of those interdependencies by competitive or collaborative strategies. For example, Feser and Sweeny (1998) use statistical procedures to identify strong inter-industry ties in input-output tables, and Rey and Matheis (2000) rely on a hybrid approach involving analysis of input–output relationships and measures of competitive strength of industries in adjacent counties. Rey and Matheis (2000, pp. 4–5) also note that many economic development practitioners use a simpler approach of measuring industry strength using location quotients, and then grouping industries into clusters based on local industry knowledge; he attributes this practical approach to Anderson (1994).

If cluster initiatives are built around clusters identified by these empirical methods, there is a risk that the initiatives will be frustrated by 'clumps' rather than actively functioning clusters in the senses identified by Porter or Rosenfeld. Whether a cluster initiative launched by local leaders will be successful may depend critically on whether firms in the region are aware of their interdependence and the nature of agglomeration economies enabled by that interdependence. Cluster initiatives launched in regions with clumps should be less successful than initiatives involving true clusters.

Industries based on scientific innovation, such as biotechnology, may be an appropriate setting for examination of the clump and cluster perspectives. Owing to the nature of the scientific innovation process, development of collaborative relationships within clusters may be an efficient and effective business strategy, in contrast to the competitive relationships Porter stresses as the key to successful clusters. Nelson (1987) emphasizes the importance of exchanges of tacit knowledge among firms engaged in research and development; these exchanges are facilitated by informal relationships among firms. Clumps, on the other hand, may be an inefficient approach to success in biotechnology and related fields. Saxenian's (1994) seminal comparison of the responses to the emergence of personal computers by computer-related firms in the Boston and Silicon Valley regions clearly illustrates the significance of the exchange of tacit knowledge among firms in a rapidly changing technological environment. The more 'clump-like' group of firms in the Boston area was much less successful in negotiating this major shift in computer technology than were firms in Silicon Valley who were distinguished by a variety of informal mechanisms for the exchange of tacit knowledge among firms. The Silicon Valley firms also competed vigorously with each other, as well as firms in other regions, demonstrating that Porter's and Rosenfeld's perspectives are not mutually incompatible.

Rickne (2004) has developed an approach to identifying the existence of clusters in industries based on scientific innovation, using data from three regions (Ohio, Massachusetts, and Sweden). Rickne suggests that life sciences evolve in an uncertain process involving research institutions, private firms, and funding organizations. Citing several authors proposing evolutionary models including Nelson (1987) and Lundvall (1992), Rickne suggests that life sciences develop in a way that favours new firm formation in confined areas. Nelson stresses the importance of tacit knowledge in some industries, knowledge that cannot be easily reduced to manuals and formulae. Individuals with experience in an industry carry tacit knowledge with them when they move from one firm to another, or in the case of biotechnology, from a university laboratory into a for-profit company.

Another factor that encourages spatial concentration in the life sciences are university rules limiting faculty roles in for-profit activities, combined with a need for a continuing relationship to university scientists by staff of for-profit firms. Part-time consulting relationships result, and the firm has an incentive to locate near the university to facilitate this interaction.

In addition, the long-term nature of the commercial opportunities in the life sciences requires large, long-term investments that are often provided by venture capital firms. In the USA, venture capitalists encourage start-up firms in which they invest to locate nearby to facilitate interaction between scientifically dominated firm personnel and venture capital staff with more expertise in finance and management. Venture capital firms usually insist on a seat on the new firm's board of directors, and they interact with the firm in a variety of other ways. Manula (2006) provides an excellent description of the roles venture capitalists play in biotechnology start-ups. Clustering in a confined geographic space serves the interests of all of these stakeholders.

Rickne also points out that in the broader field of 'bio-materials', which includes use of manufactured materials to replace various body parts, there is technological competition among competing solutions, and good ideas may come from practising physicians, university or other research organizations, private firms focused on a candidate solution, or from suppliers. Networks among the stakeholders increase the pace of discovery, favouring dense clusters with intense interactions.

Because technology is changing rapidly in the life sciences, clusters should be an efficient industry organizational solution for all of the reasons suggested above. Clumps should either be less successful or not present in this industry. Before turning to the Seattle case to examine this hypothesis, an additional important factor is considered that is not part of the cluster–clump hypothesis: the macroeconomic environment established by the global pharmaceutical industry.

2. GLOBAL AND NATIONAL BIOTECHNOLOGY DEVELOPMENT PATTERNS

Biotechnology companies have become the focus of economic development efforts in many parts of the United States (Cortright and Mayer 2002), Canada and Europe (OECD 2002, p. 58). Cities focusing on biotechnology are hoping to attract high-wage companies with strong growth prospects. For example, the City of Seattle proclaims that 'Seattle is an emerging biotech center', and 'We are committed to making Seattle a flourishing world center for biotech and the life sciences.'[2] In the San Francisco Bay area, the Economic Development Alliance for business states: 'Today, biotechnology is increasingly becoming an important economic development strategy around the world as regions and communities try to capture the economic benefits of this promising industry.'[3]

With cities and nations around the globe interested in attracting or developing biotechnology companies, it is important to have a good understanding of what drives this industry, how it relates to the international pharmaceutical industry, and what effects available policy tools may have on local concentrations of biotechnology companies.

Cortright and Mayer (2002, p. 3) find significant concentration of biotechnology industry development in nine of the 51 largest metropolitan areas in the United States. The top five metropolitan areas in Table 11.1 accounted for 75 per cent of the venture capital invested in biotechnology from 1994 to 2000, 74 per cent of the research contracts from pharmaceutical firms, and 56 per cent of the new biotechnology firms founded in those

Table 11.1 Characteristics of nine leading biotechnology areas in the United States

Metropolitan Area	Basis for industry development
Boston	Early research leader
San Francisco	Early research leader
Raleigh–Durham	Emerging centre – research institutions and many start-ups
San Diego	Emerging centre – research institutions and many start-ups
Seattle	Emerging centre – research institutions and many start-ups
New York	Pharmaceutical centre
Philadelphia	Pharmaceutical centre
Los Angeles	Home of Amgen, largest biotech
Washington DC/ Baltimore	National Institutes of Health; firms involved in human genome mapping

Source: Cortright and Mayer, 2002

years. While upwards of 200 cities may be focusing on attracting biotechnology firms, only the top five to nine metropolitan areas have achieved much success, and given the advantages established by these areas, other cities will find it hard to establish a foothold in this field.

Cockburn (2004) places these developments in a larger picture of changing structure in the pharmaceutical industry. Prior to the 1980s, basic science conducted in universities resulted in ideas for chemical combinations with positive therapeutic benefits. Further development, including medical efficacy reviews to satisfy national regulatory authorities and creation of efficient manufacturing techniques, was carried out by large pharmaceutical companies who then marketed and distributed these products. Developments in this industry prior to the 1980s resulted in large companies having a global presence, often putting their manufacturing facilities in a few low-cost locations. These corporations were vertically integrated and financed new product development with internal funds. Neither clumps nor clusters were typical arrangements in this setting; rather, large multinational corporations conducted research and manufacturing in a variety of locations based on historical evolution of each corporation and a search for least-cost manufacturing locations worldwide.

The advent of biotechnology changed these arrangements. The Bayh–Dole Act, passed by the US Congress in 1980, gave US universities a financial stake in new drug discovery based on biotechnology research. Universities began to license intellectual property developed in university laboratories as well as publishing some results in academic journals. Universities also began to encourage commercialisation activities with a variety of arrangements vis-à-vis professors and staff involved in discovering potentially valuable biotechnology formulations. The venture capital industry, formerly focused on electronic and computer products, stepped forward with financing for these new biotechnology-based start-up companies. The pharmaceutical industry reacted by either licensing products from the biotechnology companies, or by buying the companies with promising technologies. Both Cockburn (2004) and Grabowski and Wang (2006) observe that these developments were accompanied by a decrease in the filing of patents for new chemical entities for pharmaceutical uses in the 1980s and 1990s, and a shift in the focus of new filings towards biotechnology products, often for cancer treatments.

Cooke (2001) characterizes this situation rather more bluntly, noting that the major pharmaceutical companies, or 'big pharmas', are dependent on the independent biotechnology companies for new drug discovery, and at the same time, the biotechnology companies are dependent on a variety of funding arrangements with big pharmas to generate the financial resources needed to carry out the clinical trials necessary to satisfy regulatory authorities. Cooke calls this a game with asymmetrical power, since the big pharmas have extensive financial resources and the biotechnology companies cannot overcome the power asymmetry unless they themselves develop into fully integrated pharmaceutical production companies.

Sustained funding from the US government has fuelled this 'game' for many years. The FY2007 budget request from the National Cancer Institute, part of the National Institutes of Health, was for $5.9 billion.[4] For all activities under the National Institutes of Health, the budget request for FY2007 was $28.6 billion, a funding level similar to that enacted in the two previous fiscal years.[5]

In this context, what are the prospects for development of regional clusters of biotechnology companies? State Science and Technology Institute (SSTI), an association of state science and technology organizations, asked this question in a May 2006 newsletter:

> If the US biotech powerhouse areas are being reduced to mere concentrations of boutique research labs, what can other areas of the country do to be active players in biotechnology? Is being/becoming a high-wage research center for a particular niche of the biotech/life science industry a sufficient goal for (technology-based economic development) strategists?[6]

To this question about the prospects for further development of regional biotechnology clusters, one can add questions about how to analyse these entities. Is the cluster idea a useful construct for thinking through the regional development issues, or is the macroeconomic context established by the large pharmaceutical companies the dominating factor that will control development of this set of industries? This chapter examines these issues using patterns of development of biotechnology in the Seattle area.

Patterns of development in this region may shed some light on the prospects for other regions attempting to build biotechnology clusters. A description of the situation of biotechnology and related companies in the Seattle area follows; subsequent sections explore various theoretical paradigms used to understand developments in other regions, exploring the fit of these paradigms to Seattle's circumstances and examining the policy implications of the paradigms for utility and prior use in Seattle.

3. THE LIFE SCIENCES IN SEATTLE

The life sciences cluster in the Seattle area largely grew out of research programmes at the University of Washington (UW), and at a partner non-profit organization, the Fred Hutchinson Cancer Research Centers (known locally as 'the Hutch'). UW includes a medical school that developed a substantial research capacity over several decades. Appropriations of federal funds were critical in building up the capacity of this school during the tenure of Warren Magnuson, who represented the state in the US Senate from 1944 to 1980. Magnuson became a powerful figure in the appropriation process, and the appropriation bills he engineered often included funding for the UW medical school.[7] He also secured funding to establish the Hutch, transforming a family foundation into a powerful research institute that has always been closely allied with the UW medical school through joint appointments of senior research staff/faculty.

These two institutions became significant research organizations, as indicated by their success in securing competitive research grants from the National Institutes of Health, and by at least nine Nobel prizes awarded to their scientists. UW received $995 million in research funding in 2005, including $791 million in federal funds. This university is the largest recipient of federal funds among public universities.[8]

UW and the Hutch spawned a number of biotechnology start-up companies following passage of the Bayh–Dole Act in 1980. A 'genealogical chart' published by the Washington Biotechnology and Biomedical Association (WBBA) shows the linkages established by tracing the employment history of company founders from one organization to another.[9] The founders of 54 companies had some prior linkage to the University of Washington, and at least 35 company founders came out of the Hutch. Five other universities and research institutes spawned additional commercial firms, as have seven of the early commercial firms.

The growth in the research institute component of the cluster in recent years, due in no small part of philanthropic gifts by Microsoft founders Bill Gates and Paul Allen, suggests that additional for-profit start-ups are likely in the future. The philanthropic gifts and investments of these two billionaires may also be subtly changing the character and direction of the life sciences complex in the area.

An on-line database published by the WBBA provides a view of some key characteristics of this group of companies (see Table 11.2), including the specialty of each company or non-profit organization, the date of founding, and the location of the organization's main office (and usually its only physical establishment). The database lists a total of 24 research institutes, 58 companies focused on development of therapeutic products, 18 companies focused on diagnostic products (which can be biotechnology agents or physical devices), and 18 companies pursuing genomics (often with the aim of discovering therapeutic agents). This includes nonmembers as well as association members, making it the most complete listing of life sciences firms available in Washington.

Table 11.2 Characteristics of life sciences firms and organizations in Washington State

	Research Institutes	Therapeutics	Diagnostics	Genomics	Venture capital (focus on life science firms)
No of Firms	24	58	18	18	15
Estimated sector employment	12414	3706	489	670	n.a.
Founding of firms					
pre-1970	5	1			n.a.
1970–79	4	1	1		n.a.
1980–89	3	5	3		n.a.
1990–1999	5	29	6	7	n.a.
2000–06	4	22	7	11	n.a.
Unknown	3		1		
Location of firm (percentages)					
Seattle	68	55	61	61	89
Suburb	21	36	22	39	11
Different metro area in state	11	9	17	0	0

Source: Compiled by the author from the Washington Biotechnology and Biomedical Association on-line database at:http://www.wabio.com/industry/directory (July 2006)

The database also includes nine venture capital firms specializing in life sciences companies, and 15 pharmaceutical companies located in the state. The venture capital companies are mostly locally-based organizations created by investors living in the state; many of these investors have accumulated significant personal wealth from previous high-tech companies in information technology, telecommunications, or biotechnology.

The pharmaceutical companies in the area, on the other hand, include small local offices of nine major international companies including Novartis, Merck, Hoffman-LaRoche and other multinational giants of the pharmaceutical industry. All of these industry groups are significantly concentrated in the city of Seattle; 61 per cent to 89 per cent of the establishments are located inside Seattle city limits.

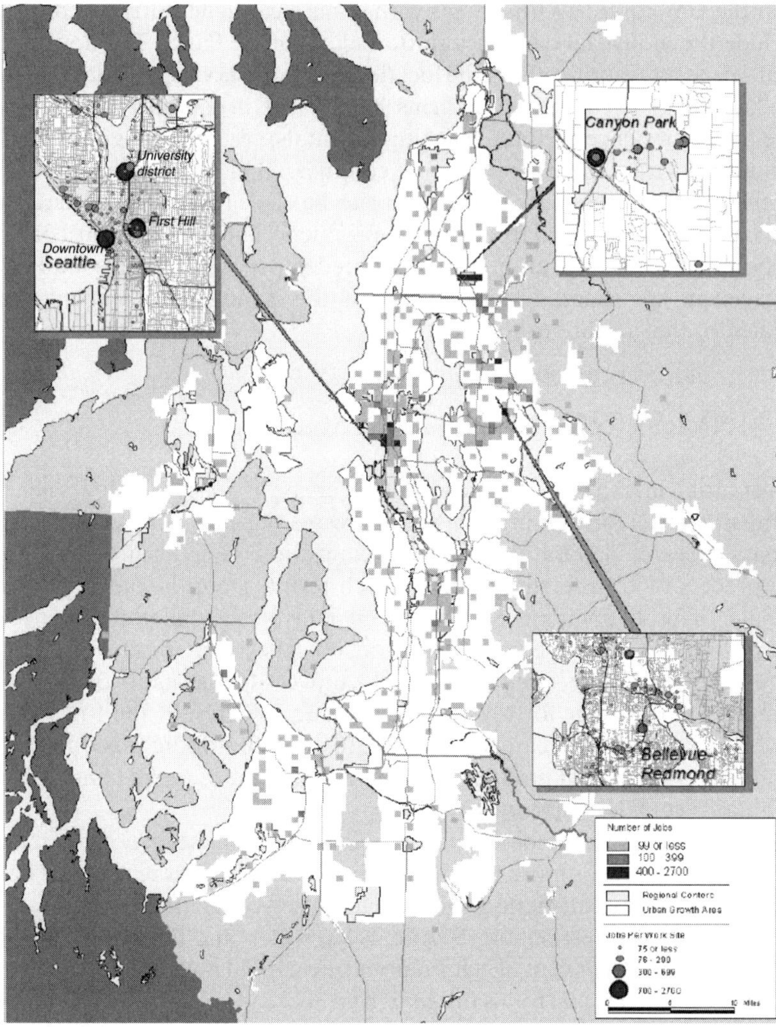

Source: *Prosperity Partnership*

Figure 11.1 Location and density of employment in the Puget Sound life sciences cluster

The life sciences firms in this region are geographically clustered as shown by Figure 11.1. While there are a few firms located throughout Seattle and its largest suburb Bellevue, the inset maps show the high concentration of firms in downtown Seattle, the First Hill neighbourhood, and near the UW campus. Other areas with a significant concentration of firms include the industrial park area in Bothell (Canyon Park). The downtown Seattle concentration reflects the location of many service providers including lawyers and venture capital firms. The First Hill concentration evolved due to the presence of three teaching hospitals; research programmes are conducted within these hospitals and commercial biotechnology firms that often chose to locate near the hospitals to be close to patients undergoing clinical trials of new pharmaceutical products. The employment concentration near the UW campus is mainly reflective of the faculty and staff engaged in life sciences research; few of the commercial companies are located in this neighbourhood.

4. INVESTING IN THE LIFE SCIENCES

Substantial investments are required to create a successful life sciences company. Ideas for new products or processes typically originate in a university research laboratory. 'Commercialization' is a term referring to the process of transferring basic research results from the university to a private for-profit company. This process can involve issuance of a licence to a private company to use technologies or substances that the university may have patented, and in some instances the university may assist the start-up company in exchange for royalties or an equity interest.[10] Ten to 15 years of development work can be required to turn promising basic research results into a potential product and then conduct the required clinical trials to obtain regulatory approval.

Consequently, most biotechnology firms are financed with a major investment by one or more venture capital firms. The Seattle area has been quite successful in attracting venture capital investments. As Figure 11.2 shows, the Northwest region (Seattle, Washington and Portland, Oregon) has attracted 3–5 per cent of national venture capital investments each year since 1995, with annual totals ranging from $0.5 to 1.5 billion. These investments have gone to over 100 firms per year in a variety of advanced technology industries. In the first quarter of 2006, 37 firms received a total of $301 million in funding, including three biotechnology firms who received a total of $21 million, and six biomedical device companies who received a total of $122 million. Venture capital investment is an area of success for Washington firms; firms in the state are receiving a higher percentage of national venture capital investment than the state's 'fair share' as indicated by population (2.1 per cent of national population).

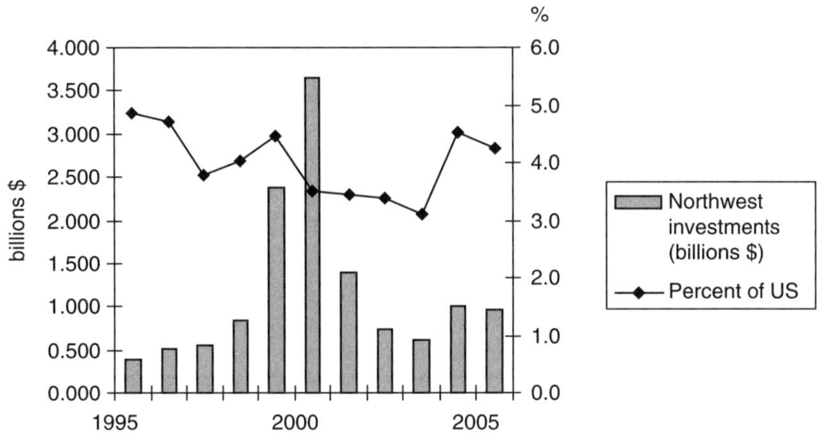

Source: PriceWaterHouseCooper 'MoneyTree'

Figure 11.2 Venture capital investments in the Northwest

5. PUBLIC SECTOR ROLES

There is a quite long history of public policy interest in promoting life science industry development in Seattle. In 1985, city, university, and research institute leaders gathered at a conference to discuss high-tech development options. They identified the South Lake Union region, located between downtown and the UW, as a likely location for biotechnology company location. Recently, major public-sector policy actions have been taken to stimulate growth of life science industries through a cluster strategy.

The Prosperity Partnership is a cluster-based economic development strategy created by the Puget Sound Regional Council (PSRC). PSRC is a multi-county agency governed by representatives of four Puget Sound counties and the many municipalities located in these counties. The principal function of the PSRC is transportation planning mandated by the federal government. However, in 2005 PSRC merged with a regional Economic Development District that disperses federal infrastructure funds and carries out planning for economic development purposes. Under a new executive director, the District board in 2004 launched a regional economic development strategy called the Prosperity Partnership.[11]

Over 600 individuals participated in drafting the strategy by attending meetings of several cluster and 'foundation' working groups. A consultant

Table 11.3 Prosperity Partnership cluster strategy

Selected Prosperity Partnership Action Items
Actions specific to life sciences: • Develop a vision for the life sciences • Create opportunities for companies and researchers to network • Ensure strong short-term results for the Life Sciences Discovery Fund.
Foundation initiatives relevant to life sciences: • Leverage the Alliance of Angels into high quality entrepreneurship mentoring • Develop an entrepreneurship agenda • Develop consensus higher education proposal.

analysis was presented at an initial public meeting in November 2004, suggesting that the four-county Puget Sound Region had significant competitive strength and strong future growth prospects in five clusters, one of which was labelled 'life sciences'. A life sciences group was one of the five cluster working groups formed subsequently, members of these groups self-identifying rather than relying on the cluster definitions used in the consultant's report. Thus, this cluster strategy is built on the simple cluster identification procedures discussed by Rey and Matheis (2000) and Anderson (1994). In addition to which foundation groups examined cross-cutting strategies such as improving higher education or tax policy.

The process used by the Prosperity Partnership allowed any working group participant to suggest an action item; votes by participants were then used to focus the initiative on a few high priority initiatives. No analysis was made of the success or failure of other cluster-based economic development initiatives as this action agenda was created. In 2005, the Prosperity Partnership published a document listing 18 specific action items (see Table 11.3 for selected action items pertaining to the life sciences cluster), including a designated organization responsible for implementing each item and a due date for completion.[12] At least a third of the items are of interest to firms in the life sciences cluster.

Interviews with two leaders in the life sciences working group provide some insight on progress in implementation of this strategy. Dr Lee Huntsman, president emeritus of the UW, indicated that the main outcome of the working group was endorsing the establishment of the Life Sciences Discovery Fund.[13] This fund was established by the state with $35 million per year in expected proceeds starting in 2008 from a major class action lawsuit against tobacco companies. The state expects these funds to be leveraged by additional private sector investments. The state legislature established a Life Sciences Discovery Fund Authority (LSDFA) to decide how to spend funds accumulated in a special account in the state treasury. Funds can be used to finance R&D projects that will improve health outcomes and lower health care costs. Projects are also expected to increase employment related to the life sciences.[14]

Dr Huntsman also indicated that the University of Washington is likely to create a new centre to provide additional capacity in the field of imaging to support private life sciences companies. State-of-the-art imaging is very expensive yet critical to life sciences companies. It is also an active research area at the university. Thus, the university is pulling together a centre concept that would provide additional laboratories for university research, but also arrangements for private-sector use of these laboratories on a contractual basis. Additional centres of this sort may be feasible for advanced instrumentation and high-performance computing.

Susannah Malarkey, executive director of the Technology Alliance, an industry association promoting the interests of advanced technology companies in Washington, indicated that there has been some effort to create a separate 'Bio Angels' group within the existing Alliance of Angels. The Alliance of Angels is an angel investor network facilitated by the Technology Alliance, its purpose being to introduce potential start-up company investment opportunities to angel investors who may wish to make equity investments. To date there has been insufficient deal flow to justify creation of a Bio Angels group. Malarkey also indicated that venture capital companies are trying to move 'upstream' into second-round financings of young companies that are expanding. Thus there may remain a gap in financing options for early-stage companies in life sciences.[15]

The Prosperity Partnership also organized a working group to deal with higher education issues. This group met several times during the spring and summer of 2006, and crafted a proposal for expanding undergraduate degree production. The Technology Alliance participated in this working group, linking it clearly to life sciences and venture capital firms who fund and serve on the board of the Technology Alliance. The higher education proposal was predicated on evidence that the state does not grant as many bachelor's degrees as a percentage of the college-age population of the state as do many other states that specialize in advanced technology industries. In granting science and engineering degrees, the state is ranked only 25th in the nation. The state's governor has separately convened public and private sector leaders to examine a variety of education issues from early childhood through public schools to college and university level offerings in the state, in an effort called 'Washington Learns'.[16] This effort resulted in proposals to the 2007 legislature, making it a good time for the Prosperity Partnership to provide additional evidence and support for initiatives that would support advanced technology clusters in the Puget Sound. In March 2007 representatives of the Prosperity Partnership, the Technology Alliance, and two other business groups presented a summary of the higher education proposal to the governing board of public higher education in the state, the Higher Education Coordinating Board. These representatives indicated that their higher education proposals, including expansion of seats for students in several scientific fields, were under active consideration by the legislature.

Thus, there seems to be some response to the Prosperity Partnership working group agenda, with movement on funding for higher education, creation of the Discovery Fund and some new initiatives at the UW that will provide private companies access to needed start-of-the art equipment in imaging and other fields. However, there is little movement on expanding access to capital or management expertise through angel investors, owing to a perception that there are too few new investment opportunities. Participation of association and education leaders in the Prosperity Partnership initiative provides support for the cluster hypothesis. Biotechnology and other life sciences firms are aware of their interdependence and work collaboratively through associations to achieve common goals.

Koehler (1996) lists factors important to start-up biotechnology companies engaged largely in research and development activities. One or more strong research universities is an essential condition; commercial companies seldom start their research and development with their own pathbreaking original research, but tend to take over promising lines of development pioneered by university researchers. Some of these academics play key roles inside commercial companies as the research transitions into product development and clinical trials. Locating the company near a research university facilitates a close relationship after the company acquires rights to a particular technology or biotech construct from the university. In addition, the universities train graduate students who may become important members of the company's workforce. Other factors needed by these firms include venture capital and other forms of financing; specialized firms dealing with risky, long-term biotechnology investments tend to locate near the research universities who supply both the intellectual capital and the staff for nearby companies.

The focus of the Prosperity Partnership on fostering close university/company relationships and on increasing the output of students trained in life science fields appears to be quite compatible with Koehler's list of significant factors. However the actual implementation of strategies by organizations affiliated with the Prosperity Partnership is uneven to date, raising some questions about the likely level of success of the economic development initiative.

6. THEORY AND LOCAL STRATEGY

With the details of the Seattle case in mind, this section returns to the clump versus cluster perspectives, examining several theoretical perspectives on regional development strategy in a globalized economy. Cluster-based development strategies suggested by several authors are considered to assess the character and power of the Seattle area strategies for the life sciences cluster.

Porter (2003) argues that strong local clusters are the key to regional economic success in a global economy. Using data from US regions, he shows that regional influence is strongly shaped by the strength and innovativeness of local industry clusters. In an earlier article, Porter (2000) applied his diamond theory to the case of regional strategy, arguing that cluster strength is determined by the interaction of four factors serving as the nodes of the diamond:

- A context for firm rivalry and strategy that encourages investment and upgrading, and is characterized by strong inter-firm competition inside the region;
- Factor input conditions, including human and physical capital, information and science/technology infrastructure;
- Related and supporting industries, including capable local suppliers and the presence of related competitive industries; and
- Demand conditions, including sophisticated and demanding customers in specialized market segments that can be served globally as well as locally.

The Seattle area life sciences cluster presents a curious mix of strengths and weaknesses viewed through the lens of Porter's diamond. There is a context for the life sciences industry that encourages investment, but this context seems to be lacking in inter-firm competition, at least with respect to product sales. The life sciences firms take advantage of local factor conditions, especially the research institutions and local venture capital. Investment in start-up companies results, but few firms have moved from R&D into manufacturing pharmaceutical products. Both biotechnology and biomedical product companies have been sold to larger corporations to gain additional capital to sustain the enterprise beyond the limits of venture capital funds. Those sales seem to cap the development of these firms. For example, Immunex, the largest company in the area as of 2002, was sold to California-based Amgen,[17] and has become a specialized research arm of Amgen rather than continuing to develop on its prior trajectory as a pharmaceutical manufacturer. Other biotechnology firms have been acquired by major pharmaceutical companies in recent years. Corixa was merged into GlaxoSmithKline, a European/US pharmaceutical giant, in March 2006.[18] In April 2006, Corus, developer of a drug for treating cystic fibrosis, received a $25 million equity investment from Gilead, an international biopharmaceutical corporation based in California, and Gilead also received the right to carry out a complete acquisition at a later date.[19] Icos, the largest independent biotech company in the area after the acquisition of Immunex, was acquired by Eli Lilly in October 2006.[20] The dilemma or 'asymmetric power game' characterized by Cooke (2001) remains very much a reality in the Seattle area. This macroeconomic factor seems to be capping the development potential of the life sciences cluster, which is effectively restricted to R&D activities by either a lack of venture capital or the irresistible lure of buyout offers from large international pharmaceutical and biomedical product companies.

Few related and supporting industries are required for an industry specialized in R&D. Few suppliers exist in the Puget Sound area for instruments or other manufactured items used in the R&D process, and competitive advantage does not stem from the use of the latest and best equipment but from the scientific capital of key employees, who typically come from a local research institute or another biotechnology company. The only related and supporting industries required are venture capital firms, lawyers specializing in intellectual property, and a few construction companies and architects capable of producing appropriate laboratory buildings. No local customers exist for the eventual products of these research companies, except for the local offices of major international pharmaceutical companies. Patients in local hospitals do provide opportunities for clinical trials. Thus, Porter's diamond does not fit this case particularly well, nor does it lead to policy or strategic ideas that either support or suggest changes in local strategies. In fact, the absence of key features of Porter's framework, both competition among the biotechnology companies in the region and the absence of key supply industries, suggest that this is not adequate as a source of ideas about how to encourage further development of this cluster.

Other studies of the clusters suggest additional policy approaches. Orsenigo's (2001) reflections on an Italian biotechnology cluster initiative that started but stalled suggests several key factors that may influence success:

- A scientific and industrial base
- Networking linking science to business
- Venture capital
- Established intellectual property rights.

The Seattle area has all of these characteristics, and the Prosperity Partnership action items are designed to improve the first two. Property rights are not a problem given the Bayh–Dole act establishing rights for university researchers funded with federal grants, and several legal firms in Seattle who specialize in intellectual property issues provide counsel, assistance in drafting contracts, and litigation services if required to protect the rights of private companies. However, industry development seems to be capped by financing needs beyond the clinical trial stage. Seattle's biotech companies are going further down the development path than Orsenigo's Italian companies, but reaching limits defined by Cooke's asymmetric power game.

Isaksen (2001) observes that transnational corporations (TNCs) are playing a larger role in many industries in the increasingly globalized economy. TNCs integrate the production process at a supranational scale, with specialized suppliers in various nations operating through subcontracts, or by being acquired by the TNC. Isaksen suggests two major strategies regions can employ to enhance their success in this context. First, regional leaders can encourage cooperation among firms by fostering the formation of innovation networks and encouraging collective learning. Thus, strategies that emphasize development clusters rather than clumps are quite important, and the Prosperity Partnership's efforts, along with those of the Technology Alliance and the Washington Biotechnology and Biomedical Association, appear exactly the right type of initiative to take in this context. Second, Isaksen suggests that regions can bind units of TNCs to the region by upgrading local knowledge organizations, stimulating collective learning, and creating collective competencies. How successful this strategy has been in the case of the Seattle region is not clear. The large pharmaceutical companies with offices in the region have not participated openly in the life science associations or the Prosperity Partnership working group. The director of the WBBA has little contact with these firms and professes little knowledge of their goals or strategies in the region. One major biotechnology firm that acquired a local firm left much of the leadership of the acquired firm in place, allowing these leaders to continue to participate in some of the local associations and collaborative activities. At least in this case, a degree of TNC 'binding' has taken place.

The pharmaceutical TNCs, relying on biotechnology clusters in various regions, appear to fit Isaksen's diagnosis quite well. Cockburn (2004) documents the global nature of pharmaceutical corporations, and their increasing reliance on independent biotechnology companies as R&D suppliers. Zeller (2001) observes similar phenomena at work in Germany, and suggests that there is a tension around transmission of tacit knowledge among research companies. This tension preserves a role for regional agglomerations in biotechnology that ultimately must finance themselves by developing relationships with international pharmaceutical corporations. Zeller's description of German biotechnology fits the Seattle case well. Agglomeration certainly exists, and mechanisms exist for transmission of tacit knowledge, including venture capital firm involvement with the firms they finance, and research institute seminars that key staff from private companies attend. In addition, two industry associations provide venues for key staff members of firms to meet and network.

How does the Seattle area's strategic response compare to Isaksen's policy prescriptions? The Prosperity Partnership strategy of increasing baccalaureate degree production and supporting the implementation of the Life Sciences Discovery Fund fit well with Isaksen's second strategy idea, binding pharmaceutical TNCs to the area, although the TNCs themselves are not typically direct participants in collaborative efforts. The WBBA and Technology Alliance can also be interpreted as exemplars of fostering innovation networks and collective learning. Thus, this area's strategy seems to be congruent with Isaksen's policy prescriptions.

The genealogical chart published by the WBBA[21] suggests additional mechanisms are at work in the Seattle area, favouring the cluster hypothesis based on exchange of tacit knowledge as advanced by Rosenfeld, Nelson, Rickne and others. Many of the firms in the Seattle area trace their origins back to basic research programmes at the UW, the Hutch, or another local research institute. The chart also shows a second round of start-ups emerging from the first round in a number of cases. The 'tacit knowledge' Nelson (1987) discusses may be a key mechanism that ties new firms to the region where the original basic research was conducted, giving an advantage to regions with major research institutes.

Gilbert and Kusar (2006) also conclude that knowledge spillovers within a regional cluster are important to firms engaged in new product development. In the Seattle case, it is likely that spillovers are occurring. The executive director of the WBBA indicates that few for-profit life sciences firms in the area are direct competitors, and that there is a significant amount of collaborative activity due in part to the absence of competitive relationships. For example, the Prosperity Partnership initiative discussed below provided a new venue for collaborative action.[22]

In terms of the 'clump or cluster' dichotomy, considerable evidence exists for a cluster in the Seattle area based on research institution/biotechnology company relationships, the exchange of tacit knowledge across the university/company boundaries, and firm location within a confined region. Rickne's key hypotheses are confirmed for the Seattle case.

7. POLICY OPTIONS

The State Science and Technology Institute has posed the key question for regional policy-makers interested in enhancing the prospects for biotechnology in a particular region: is a boutique research cluster with high wages a sufficient goal? This section explores possible policy options to encourage development of biotechnology firms into fully integrated pharmaceutical companies (FIPCOs). Growing FIPCOs from the existing biotech companies is the logical alternative to continuing Cooke's asymmetric power game, which ends with turning over the fruits of local R&D to pharmaceutical TNCs. The discussion is tailored towards the situation in the Seattle area, but many of the options have relevance to other regions with similar goals.

If the development of FIPCOs is the goal, massive equity investments will be required to create the manufacturing, marketing, and distribution capabilities. How will the necessary capital be supplied?

One scenario is a new round of venture capital investment based on promising clinical trial results. R&D firms would add manufacturing capacity and use subsequent retained earnings to sustain the firm. Public policy can play a role in encouraging this outcome with tax incentives for manufacturing investments.

Another strategy is to build alliances between existing contract drug manufacturing firms and biotechnology companies with a promising product line. At present, there are several contract manufacturing firms in Washington who produce modest quantities of a new drug on contract to supply clinical trial operations. A strategic alliance between one of these firms and a biotechnology firm could instantly create a potential FIPCO without a massive infusion of new capital. Such alliance could be encouraged with tax incentives, perhaps deferring or forgiving the state's gross receipts tax for several years to allow the combined firm to establish a foothold in the market.

A third option would be to provide guidance to public research institutions regarding licensing. Rather than allowing these institutions just to maximize their revenues from commercialization activities, new policies could establish a preference for an in-state licensee, or specify a substantially higher fee for a licensee who takes the technology out of the state. Alternatively, the state could establish a tax preference for firms that create in-state manufacturing for a technology licensed from a public research institution.

A fourth policy option is to abandon the goal of fostering FIPCOs and focus on building up the strength of the existing biotechnology cluster. In essence this is the current policy, but it could be strengthened by suggestions from the Prosperity Partnership and others that university-level education programmes be strengthened, that quality of life investments be made to maintain or enhance the attractiveness of the area to talented scientists who might be recruited by private firms and research institutions, and that existing capital sources be augmented with public funds.

A final policy option is a variant on the previous research-focused strategy, but recognizes changes in the nature of the research effort reflecting the complete mapping of the human genome, the emergence of newer research fields such as nanotechnology, genomics, etc., and the local influence of the Bill and Melinda Gates Foundation. Rather than focusing solely on biotechnology research, the Seattle area could adopt the goal of a broader life science research and development cluster. New centres of excellence could be created within research institutions, with careful attention to commercialization opportunities that may emerge from these new research activities.

This is in fact an emerging strategy in Washington, with a nanotechnology centre created at the University of Washington in 1997,[23] and a joint centre created by Washington State and the Pacific Northwest National Laboratory in eastern Washington in 2001.[24] A new imaging laboratory is being designed at the University of Washington and may be launched shortly.

Local leaders have stopped talking about the biotechnology industry, emphasizing instead the term 'life sciences cluster' to reflect the broadening of the research enterprise and an implicit goal of encouraging a broader range of commercial spin-offs. The state or regional economic development organizations could establish metrics to monitor progress on these initiatives, perhaps tracking relevant research funding, patent filings, and related company formation trends.

A key feature of this strategy is recognition that the area's competitive advantage is in research, not in pharmaceutical product manufacturing. Rather than focusing incentives on manufacturing as the state has done in the past, it would shift to focusing on getting technologies to the point where an acquisition by a pharmaceutical TNC is possible, and then encourage re-cycling of the proceeds from the sale of a technology or a start-up company into new start-ups in the region based on the next round of research emerging from the research institutes.

Unfortunately the State of Washington does not tax profits or capital gains, so it has no convenient policy option to establish an incentive for recycling the wealth gain coming from the sale of a start-up company. Or to put positive 'spin' on the situation, perhaps it is the absence of income taxes in Washington that encourages the degree of wealth recycling that exists in the state.

There is an aphorism mentioned by many economic development practitioners that 'all economic development is local'. This policy options section suggests those that are tailored to the particular circumstances of the Seattle region. A similar process of developing tailored regional options will be necessary in other regions promoting life science-based industries since the generic prescriptions emerging from the cluster strategy and research milieu research traditions are well known and already applied in many parts of the world. To create new regional competitive advantages in this context requires a fresh start based on existing local conditions as well as insights about future development options for each region.

NOTES

1. Conversation with Bob Drewel, Executive Director, Puget Sound Regional Council, August 2004.
2. See City of Seattle website for economic development at: http://www.seattle.gov/economicdevelopment/pages/maj_ind_sectors.htm#Anchor-Biotec-64703, August 2006.
3. Economic Development Alliance for Business, 'A critical analysis of the biotechnology industry cluster in Alameda, Contra Costa, and Solano Counties', http://www.edab.org/study/East%20Bay%20Biotechnology%20Study%20Abstract.pdf, August 2006.
4. http://plan.cancer.gov/NCIBudgetRequest.shtml, August 2006.
5. http://officeofbudget.od.nih.gov/pdf/Press%20info%20final.pdf, August 2006.
6. 'Foreign bias for location in partnering with U.S. biotech start-ups', SSTI May 15 2006 http://www.ssti.org/Digest/2006/headlines 06.htm, July 2006.
7. Former Vice President Walter Mondale quipped about Senator Magnuson 'He is scrupulously fair with federal funds; one half for Washington state, one half for the rest of the country': http://www.historylink.org/essays/output.cfm?file_id=5569, July 2006.
8. http://www.washington.edu/research/statistics.html
9. See http://www.wabio.com/biohistory/index_html, 2006.
10. Seredigm, a new biotechnology start-up company, is reported to be partnering with two University of Washington scientists in its initial R&D efforts and the university will receive an equity stake in the company in return. See *Seattle Post-Intelligencer* 'Venture Capital: Major VCs stoked on startup's new drug', http://seattlepi.nwsource.com/venture/279207_vc28.html, July 2006.
11. See http://www.prosperitypartnership.org/, July 2006.
12. http://www.prosperitypartnership.org/strategy/actionitems.pdf, July 2006.
13. Interview with Dr Lee Huntsman, president emeritus, University of Washington, July 26, 2006.
14. See a description of this Authority at http://www.hecb.wa.gov/boardmtgs/documents/5-sept22-05.LifeSciencesDiscoveryFundAuthority.pdf, July 2006.
15. Interview with Susannah Malarkey, Executive Director, Technology Alliance, 26 July 2006.
16. See Washington Learns website at: http://www.washingtonlearns.wa.gov/ (July 2006).
17. http://my.amgen.com/pdfs/immunex/pressRelease020429.pdf, October 2006.
18. http://en.wikipedia.org/wiki/Corixa, August 2006.
19. http://www.coruspharma.com/home/pressrelease_12apr 06.pdf, August 2006.
20. http://newsroom.lilly.com/ReleaseDetail.cfm?ReleaseID=214900, 2006.
21. cf. note 9.
22. Interview with Jack Faris, Executive Director, Washington Biotechnology and Biomedical Association, August 3, 2006.

23. http://www.nano.washington.edu/index.asp, August 2006.
24. http://www.nano.washington.edu/pnnl/overview.html, August 2006.

REFERENCES

Anderson, G. (1994), 'Industry clustering for economic development', *Economic Development Review*, **12** (2), 26–32.

Cockburn, I. (2004), 'The changing structure of the pharmaceutical industry', *Health Affairs*, **21** (1), 10–22.

Cooke, P. (2001), 'Biotechnology clusters in the U.K.: lessons from localization in the commercialization of science', *Small Business Economics*, **17** (1–2), 43–59.

Cortright, J. and H. Mayer (2002), *Signs of Life: The Growth of Biotechnology Centers in the U.S.*, Washington, DC: Brookings Institution.

Feser, E.J. and S.H. Sweeney (1998), 'A test for spatio-economic clustering', University of North Carolina, working paper, Chapel Hill, NC.

Gilbert, B. and M. Kusar (2006), 'The influence of geographic clusters and knowledge spillovers on the product innovation activities of new ventures', in *Papers on Entrepreneurship, Growth and Public Policy*, Jena: Max Planck Institute of Economics.

Grabowski, H.B. and Y.R. Wang (2006), 'The quantity and quality of worldwide new drug introductions, 1982–2003', *Health Affairs*, **25** (2), 452–60.

Isaksen, A. (2001), 'Building regional innovation systems: is endogenous industrial development possible in the global economy?', *Canadian Journal of Regional Science*, **24** (1), 101–20.

Isard, W. (ed.) (1998), *Methods of Interregional and Regional Analysis*, Aldershot: Ashgate.

Koehler, G. (1996), 'Bioindustry: a description of California's bioindustry and summary of the public issues affecting its development', Sacramento, CA: California Research Bureau, accessed July 2006 at www.library.ca.gov/CRB/96/07/index.html.

Lundvall, B.-A. (1992), *National Systems of Innovation: Toward a Theory of Innovation and Interactive Learning*, London: Pinter.

Manula, M. (2006), *The Perceived Value-added of Venture Capital Investors: Evidence from Finnish Biotechnology Industry*, Helsinki: Research Institute of Finnish Economy.

Nelson, R. (1987), *Understanding Technical Change as an Evolutionary Process*, Amsterdam: Elsevier.

OECD (2002), *OECD Science and Technology Outlook*, Paris: OECD.

Orsenigo, L. (2001), 'The (failed) development of a biotechnology cluster: the case of Lombardy', *Small Business Economics*, **17**, 77–92.

Porter M.E. (1990), *The Competitive Advantage of Nations*, New York: Basic Books.

Porter M.E. (2000), 'Location, competition, and economic development: local clusters in a global economy', *Economic Development Quarterly*, **14** (1), 15–34.

Porter M.E. (2003), 'The economic performance of regions', *Regional Studies*, **37** (6–7), 549–78.

Rey, S. and D. Matheis (2000), *Identifying Regional Clusters in California: Vol. I. Conceptual Design*, report for California Employment Development Department, San Diego, CA: San Diego State University Department of Geography.

Rickne, A. (2004), 'Prefirm activities in the life science sector', *Environment and Planning C: Government and Policy*, **22**, 249–69.
Rosenfeld, S. (2005), Industry clusters: business choice, policy outcome, or branding strategy? *Journal of New Business Ideas and Trends*, **3** (2), 4–13.
Saxenian, A.L. (1994), *Regional Advantage: Culture and Competition in Silicon Valley and Route 128*, Cambridge, MA: Harvard University Press.
Zeller, C. (2001), 'Clustering biotech: a recipe for success? Spatial patterns of growth of biotechnology in Munich, Rhineland and Hamburg', *Small Business Economics*, **17**, 123–41.

Index